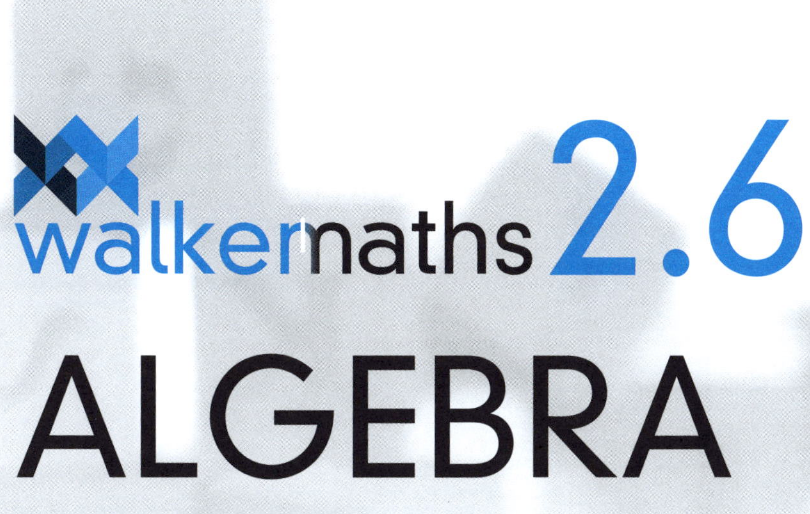

walkernaths 2.6
ALGEBRA

NCEA Level 2 External

Charlotte Walker and Victoria Walker

Walker Maths 2.6 Algebra
1st Edition
Charlotte Walker
Victoria Walker

Editor: Eva Chan
Cover and text design: Cheryl Smith, Macarn Design
Production controller: Siew Han Ong

Any URLs contained in this publication were checked for currency during the production process. Note, however, that the publisher cannot vouch for the ongoing currency of URLs.

Acknowledgements
Cover photo courtesy of Shutterstock.

We wish to thank the Boards of Trustees of Darfield and Riccarton High Schools for allowing us to use materials and ideas developed while teaching. Our thanks also go to all past and present colleagues who have generously shared their expertise and ideas.

For product information and technology assistance,
in Australia call **1300 790 853**;
in New Zealand call **0800 449 725**

For permission to use material from this text or product, please email **aust.permissions@cengage.com**

National Library of New Zealand Cataloguing-in-Publication Data
A catalogue record for this book is available from the National Library of New Zealand.

ISBN 978 0 17 035422 6

Cengage Learning Australia
Level 7, 80 Dorcas Street
South Melbourne, Victoria Australia 3205

Cengage Learning New Zealand
Unit 4B Rosedale Office Park
331 Rosedale Road, Albany, North Shore 0632, NZ

For learning solutions, visit **cengage.co.nz**

Printed in China by 1010 Printing International Limited
20 25

CONTENTS

ISBN: 9780170354226

Formulae

These are the algebra formulae which are supplied in the external examination.

Quadratic Equations	If $ax^2 + bx + c = 0$ then $x = \dfrac{-b \pm \sqrt{b^2 - 4ac}}{2a}$ and $\Delta = b^2 - 4ac$
Logarithms	If $y = b^x$ then $x = \log_b y$ $\log_b(xy) = \log_b(x) + \log_b(y)$ $\log_b\left(\dfrac{x}{y}\right) = \log_b(x) - \log_b(y)$ $\log_b(x^n) = n\log_b x$ If $y = e^x$ then $x = \log_e y$ (= $\ln y$)

 ISBN: 9780170354226

Glossary

Make your own glossary of key terms:

Term	Definition	Picture/Example
Solve		
Simplify		
Evaluate		
Power		
Index (plural: indices)		
Denominator		
Numerator		
Exponent		
Variable		
Surd		

ISBN: 9780170354226

Term	Definition	Picture/Example
Coefficient		
Constant		
Quadratic		
Expand		
Factorise		
Logarithm (log)		
Root		
Discriminant		
Appreciation		
Depreciation		

ISBN: 9780170354226

Powers

The words **power**, **exponent** and **index** all mean the same thing.

coefficient → $3a^5$ ← power, exponent, index

base

Remember:

- $a^4 = a \times a \times a \times a$
- $a^1 = a$
- $a^0 = 1$ — As long as $a \neq 0$

Multiplying powers

When multiplying terms, we **add** the indices.

Examples:

1 $a^6 \times a^3 = a^9$ — $6 + 3 = 9$

2 $2a^4 \times 6a^7 = 12a^{11}$ — Always deal with the **coefficients** first

Simplify these.

1 $b^2 \times b^3$

2 $y^8 \times y^6$

3 $15b^{10} \times 2b^5$

4 $y^3 \times y \times y^4$

5 $2z^2 \times 4z^2 \times z^6$

6 $y^{0.5} \times y$

7 $-2y^4 \times 3y^2$

8 $y^{2.5} \times y^{3.1}$

9 $-9y^5 \times (-y^2)$

10 $5y^{0.4} \times (-3y^2)$

ISBN: 9780170354226

Dividing powers

When dividing terms, we **subtract** the indices.

Examples:

1 $a^6 \div a^2 = a^4$ or $\frac{a^6}{a^2} = a^4$ — $6 - 2 = 4$

2 $12a^2 \div 6a^7 = 2a^{-5}$ or $\frac{2}{a^5}$ — Always deal with the **coefficients** first

Alternatively: $\frac{12a^2}{6a^7} = \frac{2}{a^5}$ or $2a^{-5}$

Simplify these.

1 $b^{12} \div b^2$

2 $\frac{y^8}{y^6}$

3 $\frac{y^4}{y^7}$

4 $\frac{10b^{10}}{2b^5}$

5 $\frac{15y^4}{30y^9}$

6 $\frac{x^2y^4}{y^3z}$

7 $\frac{y^4}{3y^6z}$

8 $\frac{12xy^2z}{4y^3z}$

9 $\frac{-6x^4}{2x^3}$

10 $\frac{-10x^4z^3}{-5x^2y^3z}$

ISBN: 9780170354226

Powers of powers

When finding a power of a term with a power, we **multiply** the indices.

Examples:

1 $(a^6)^3 = a^6 \times a^6 \times a^6$
$= a^{18}$ — 6 x 3 = 18

2 $(4a^3)^2 = 4a^3 \times 4a^3$
$= 16a^6$

3 $\left(\frac{1}{2}a^4\right)^3 = \left(\frac{1}{2}\right)^3 \times (a^4)^3$
$= \frac{1}{8}a^{12}$ or $\frac{a^{12}}{8}$

4 $\left(-\frac{2}{3}a^2b^3\right)^3 = \left(-\frac{2}{3}\right)^3 \times (a^2)^3 \times (b^3)^3$
$= -\frac{8}{27}a^6b^9$ or $\frac{-8a^6b^9}{27}$

Simplify these.

1 $(b^{12})^2$

2 $(2x^6)^4$

3 $(-a^7)^2$

4 $(-y^3)^5$

5 $(-3x^5)^2$

6 $(-4x^4)^3$

7 $-2(5x^3y^4)^2$

8 $(6x^2y^4z^3)^3$

9 $-(-4a^3 2c^4 b^2)^2$

10 $2(-3xy^3z^2)^2$

11 $\left(\frac{1}{5}x^2\right)^3$

12 $\left(-\frac{3}{8}x^4\right)^2$

ISBN: 9780170354226

Putting it together so far

Examples:

1 $\frac{a^3 \times a^5}{a^4 \times a^2} = \frac{a^8}{a^6}$

$= a^2$

2 $\frac{(4ab^3)^3}{2a^2} = \frac{64a^3b^9}{2a^2}$

$= 32ab^9$

3 $\frac{(2a^2)^3}{(3a^4)^2} = \frac{8a^6}{9a^8}$

$= \frac{8}{9a^2}$

4 $(a^4)^2\ (2a)^3 = a^8 \times 8 \times a^3$

$= 8a^{11}$

Simplify these.

1 $b^{12} \times b^2 \div b^4$

2 $\frac{x^2y^2z^2}{(xyz)^2}$

3 $\frac{c^3b \times cb^2}{cb}$

4 $\frac{(5b)^2}{4b^3}$

5 $\left(\frac{a^2}{2}\right)^3$

6 $\left(\frac{4z^{0.5}}{7}\right)^2$

7 $\left(\frac{6z^{0.5}}{2z}\right)^3$

8 $\left(\frac{2z^3y^2}{5xy^4}\right)^2$

9 $\left(\frac{-2a\,b^3}{b^5}\right)^2$

10 $\left(\frac{-a^3b}{4ab^2}\right)^3$

 ISBN: 9780170354226

Negative powers

1 With numbers

$$x^{-a} = \frac{1}{x^a}$$

Examples:

1 $7^{-1} = \frac{1}{7}$

2 $\left(\frac{7}{8}\right)^{-1} = \frac{8}{7}$

3 $\left(\frac{3}{4}\right)^{-2} = \frac{4^2}{3^2}$

$= \frac{16}{9}$

4 $\left(1\frac{2}{3}\right)^{-2} = \left(\frac{5}{3}\right)^{-2}$

$= \frac{9}{25}$

5 $(-2)^{-4} = \frac{1}{(-2)^4}$

$= \frac{1}{16}$

6 $(-2)^{-3} = \frac{1}{(-2)^3}$

$= -\frac{1}{8}$

Simplify these.

1 3^{-2}

2 $\left(\frac{4}{5}\right)^{-1}$

3 0.25^{-1}

4 $\left(\frac{2}{3}\right)^{-3}$

5 $\left(\frac{9}{5}\right)^{-1}$

6 $\left(\frac{4}{3}\right)^{-2}$

7 $\left(1\frac{1}{4}\right)^{-2}$

8 2.5^{-3}

9 $\left(-\frac{1}{2}\right)^{-4}$

10 $(-3)^{-4}$

11 $(-4)^{-3}$

12 $\left(-\frac{7}{5}\right)^{-2}$

ISBN: 9780170354226

2 With variables

$$x^{-a} = \frac{1}{x^a}$$

Examples:

1 $a^{-3} = \frac{1}{a^3}$

2 $2a^{-1}b^{-4}c^3 = \frac{2c^3}{ab^4}$

3 $3y^{-2} = 3 \times \frac{1}{y^2}$

$= \frac{3}{y^2}$

4 $(ab^3)^{-5} = a^{-5} \times b^{-15}$

$= \frac{1}{a^5 b^{15}}$

Simplify these.

1 b^{-12}

2 $5a^{-2}$

3 ab^{-2}

4 $8a^4 b^{-3}$

5 $(a^2 b^3)^{-1}$

6 $(3ab^5)^{-2}$

7 $2(a^2 b^4 c)^{-3}$

8 $\left(\frac{1}{a}\right)^{-3}$

9 $\left(\frac{1}{a^5 b}\right)^{-2}$

10 $3\left(\frac{c^2}{a^2 b^4}\right)^{-1}$

11 $\left(\frac{5}{a^{-2} b}\right)^{-2}$

12 $\left(\frac{4c^3}{a^2 b^{-1}}\right)^{-1}$

ISBN: 9780170354226

Fractional powers — surds

- Terms with fractional powers can be written in surd form.
- A term in surd form has a root sign ($\sqrt{\ }$).
- Remember that $\sqrt{\ }$ means $\sqrt[2]{\ }$, the **square** root.

index form — $x^{\frac{3}{5}} = \sqrt[5]{x^3}$ — surd form → contains a $\sqrt{\ }$ sign

Examples:

Index form		Surd form
$a^{\frac{1}{3}}$	=	$\sqrt[3]{a}$
$a^{\frac{7}{2}}$	=	$\sqrt{a^7}$
$7x^{\frac{1}{5}}$	=	$7\sqrt[5]{x}$

Write these in index form.

1 $\sqrt[6]{a}$

2 $\sqrt[4]{a^3}$

3 $\sqrt{a}$

4 $\sqrt{a^{16}}$

5 $9\sqrt[5]{a^{10}}$

6 $3\sqrt[4]{a^5}$

Write these in surd form.

1 $a^{\frac{1}{3}}$

2 $a^{\frac{2}{5}}$

3 $8a^{\frac{1}{3}}$

4 $(8a)^{\frac{1}{3}}$

5 $12a^{\frac{5}{7}}$

6 $9a^{\frac{3}{2}}$

ISBN: 9780170354226

Calculations involving surds:

1 Convert the expression into index form.

2 Use the power laws to simplify the expression.

Examples:

1 $\sqrt{a^5} \times \sqrt{a^3} = a^{\frac{5}{2}} \times a^{\frac{3}{2}}$

$= a^{\frac{8}{2}}$

$= a^4$

2 $\sqrt[3]{a^7} \times a^2 = a^{\frac{7}{3}} \times a^2$

$= a^{\frac{13}{3}}$

3 $\dfrac{\sqrt{a^5}}{\sqrt{a^3}} = \dfrac{a^{\frac{5}{2}}}{a^{\frac{3}{2}}}$

$= a^{\frac{5}{2}-\frac{3}{2}}$

$= a^{\frac{2}{2}}$

$= a$

4 $\dfrac{\sqrt{a^8}}{\sqrt{a^6}} = \dfrac{a^4}{a^{\frac{6}{2}}}$

$= \dfrac{a^4}{a^3}$

$= a$

Simplify or evaluate the following.

1 $\sqrt{a^3} \times \sqrt{a^7}$

2 $\sqrt[3]{a^4} \times \sqrt[3]{a^5}$

3 $\sqrt[4]{a^3} \times \sqrt{a^3}$

4 $\sqrt[5]{a^3} \times a$

5 $\sqrt[3]{a^6} \times a^4$

6 $\dfrac{\sqrt[4]{a^5}}{\sqrt[4]{a^3}}$

7 $\dfrac{\sqrt[8]{a^5}}{\sqrt[4]{a}}$

8 $\dfrac{a^2}{\sqrt[5]{a^4}}$

ISBN: 9780170354226

Combined roots, powers and coefficients

1. Take the required root of the **coefficient**

$$\sqrt[3]{8y^5} = \sqrt[3]{8} \times y^{\frac{5}{3}} = 2y^{\frac{5}{3}}$$

2. Divide the index (5) by the size of the root (3)

Examples:

1 $11\sqrt{a^5} = 11a^{\frac{5}{2}}$

2 $\sqrt{49a^3} = 7(a^3)^{\frac{1}{2}} = 7a^{\frac{3}{2}}$

3 $\sqrt[5]{32a^{10}} = 32^{\frac{1}{5}}(a^{10})^{\frac{1}{5}} = 2a^2$

4 $\sqrt[3]{\frac{8}{27}a^{12}} = \frac{2}{3}a^{\frac{12}{3}} = \frac{2}{3}a^4$ or $\frac{2a^4}{3}$

Simplify these.

1 $\sqrt[4]{16a^8}$

2 $\sqrt[3]{64a^9}$

3 $9\sqrt[5]{a^3}$

4 $7\sqrt{a^{11}}$

5 $(64a)^{\frac{1}{3}}$

6 $(25a^4)^{\frac{1}{2}} \times (8a^6)^{\frac{1}{3}}$

7 $\left(\frac{4}{25}x^{\frac{1}{3}}\right)^{\frac{1}{2}}$

8 $(8x^4)^{\frac{2}{3}} \times (36x^{\frac{2}{3}})^{\frac{1}{2}}$

ISBN: 9780170354226

Powers where the base is a number

1 Rewrite the expression with the smallest base possible.
2 Use the power laws to simplify the expression.

Remember: 2^{n+1} can be written as $2^n \times 2^1$, which is $2(2^n)$.

Examples:

1 $8^x = (2^3)^x$
$= 2^{3x}$

2 $81^{\frac{n}{2}} = (3^4)^{\frac{n}{2}}$
$= 3^{2n}$

3 $\frac{9^{2n}}{3^{4n}} = \frac{(3^2)^{2n}}{3^{4n}}$
$= \frac{3^{4n}}{3^{4n}}$
$= 1$

4 $\frac{2^{n+1}}{8} = \frac{2^{n+1}}{2^3}$
$= 2^{(n+1)-3}$
$= 2^{n-2}$

Simplify these.

1 $\frac{2^{3n}}{8}$

2 $\frac{27}{3^{2n}}$

3 $\frac{64^n}{2^{5n}}$

4 $\frac{2^n \times 4^{3n}}{16^{2n}}$

5 $\frac{3 \times 4^{5n}}{2^n}$

6 $\frac{2^{n+3}}{16}$

7 $\frac{3^{2n-1}}{9^n}$

8 $\frac{8^{n+1} \times 4^n}{2^{3n-1}}$

ISBN: 9780170354226

Powers — putting it all together

Example 1:

$$\left(\frac{4x^8}{25x^2}\right)^{-\frac{1}{2}} = \left(\frac{25x^2}{4x^8}\right)^{\frac{1}{2}}$$
$$= \frac{5x}{2x^4}$$
$$= \frac{5}{2x^3}$$

Example 2:

$$\sqrt{(125x^9)^{-\frac{2}{3}}} = \sqrt{\frac{1}{(125x^9)^{\frac{2}{3}}}}$$
$$= \sqrt{\frac{1}{25x^6}}$$
$$= \frac{1}{5x^3}$$

Simplify these.

1 $(3x^4)^2(x^2)^3$

2 $\dfrac{(2x^3)^4}{\sqrt{16x^2}}$

3 $2(27x^6)^{\frac{1}{3}}$

4 $(9x^{10})^{-\frac{1}{2}}$

5 $7(25x^4)^{-\frac{1}{2}}$

6 $\left(\dfrac{x^{15}}{27x^9}\right)^{-\frac{1}{3}}$

7 $\left(\dfrac{1}{64x^{15}}\right)^{-\frac{2}{3}}$

8 $3(16x^{12})^{-\frac{3}{4}}$

ISBN: 9780170354226

9 $\dfrac{(8x^6)^{\frac{1}{3}}}{2x^4}$

10 $(5x^3)\,(25x^4)^{-\frac{1}{2}}$

11 $\sqrt{\dfrac{4x^8y^2}{x^{-2}}}$

12 $(4a^5 \times 25a^3)^{-\frac{1}{2}}$

13 $(8x^6)^{-\frac{2}{3}}$

14 $\left(\dfrac{4x^2}{x^4y^6}\right)^{-\frac{1}{2}}$

15 $\dfrac{(16x^{12})^{-\frac{3}{4}}}{(25x^{10})^{-\frac{1}{2}}}$

16 $\sqrt{(64x^{60})^{-\frac{1}{3}}}$

17 $\sqrt{\left(\dfrac{64}{x^{18}y^{12}}\right)^{-\frac{1}{3}}}$

18 $\sqrt{4\left(\dfrac{36y^6}{x^8}\right)^{-1}}$

ISBN: 9780170354226

Logarithms

Logarithm introduction

Logarithms (log) and exponents are very closely related:

$b^x = y \leftrightarrow \log_b y = x$

This is given to you on the formula sheet.

Notice that the **power** (exponent) to which the base is raised is the same as the value of the **log**.

	Exponent form		Log form
Examples:	$10^2 = 100$	$\leftrightarrow$	$\log_{10} 100 = 2$
	$3^4 = 81$	$\leftrightarrow$	$\log_3 81 = 4$
	$2^7 = 128$	$\leftrightarrow$	$\log_2 128 = 7$
Note:	$10^3 = 1000$	$\leftrightarrow$	$\log_{10} 1000 = 3$ or $\log_{10} 10^3 = 3$

If **these** match, the answer is the power.

Write these in log form.

1 $2^5 = 32$

2 $10^4 = 10\,000$

3 $a^b = c$

4 $10^1 = 10$

5 $5^3 = 125$

6 $3^4 = 81$

Write these in exponent form.

1 $\log_4 16 = 2$

2 $\log_{10} 100 = 2$

3 $\log_2 64 = 6$

4 $\log_5 625 = 4$

5 $\log_6 36 = 2$

6 $\log_3 27 = 3$

ISBN: 9780170354226

Some tricks:

1 $\log_b 1 = 0$ because $(\text{anything})^0 = 1$

2 $\log_b b = 1$ because $b = b^1$

3 $\log_b \frac{1}{b^x} = -x$ because $\frac{1}{b^x} = b^{-x}$

4 $\log_b(\sqrt[n]{b^m}) = \frac{m}{n}$ because $\sqrt[n]{b^m} = b^{\frac{m}{n}}$

Examples:

Exponent form	Log form
$10^0 = 1$	$\log_{10} 1 = 0$
$23^1 = 23$	$\log_{23} 23 = 1$
$\frac{1}{32} = \frac{1}{2^5} = 2^{-5}$	$\log_2 \frac{1}{32} = \log_2 2^{-5} = -5$
$4 = \sqrt[3]{8^2} = 8^{\frac{2}{3}}$	$\log_8 4 = \log_8(8^{\frac{2}{3}}) = \frac{2}{3}$

Write these in log form.

1 $13^0 = 1$

2 $5^1 = 5$

3 $10^{-3} = 0.001$

4 $9^{\frac{1}{2}} = 3$

5 $(4)^{-1} = \frac{1}{4}$

6 $10^{-2} = 0.01$

7 $16^{-\frac{1}{2}} = \frac{1}{4}$

8 $25^{0.5} = 5$

9 $125^{\frac{1}{3}} = 5$

10 $8^{-\frac{1}{3}} = \frac{1}{2}$

Write these in exponent form.

1 $\log_{15} 1 = 0$

2 $\log_{10} \frac{1}{1000} = -3$

3 $\log_{36} 6 = \frac{1}{2}$

4 $\log_{16} 16 = 1$

5 $\log_{32} 2 = \frac{1}{5}$

6 $\log_{10} 0.1 = -1$

7 $\log_8 1 = 0$

8 $\log_2 \frac{1}{2} = -1$

9 $\log_{49} \frac{1}{7} = -\frac{1}{2}$

10 $\log_{1000} 0.1 = -\frac{1}{3}$

11 Match each log statement with its correct value.

Log	Value
$\log_2 200$	0.25
$\log_{10} 4000$	4.192
$\log_5 1.5$	3.602
$\log_7 300$	5.1404
$\log_{100} 240$	1.19
$\log_6 10\,000$	6.251
$\log_3 100$	7.645
$\log_3 960$	2.93

ISBN: 9780170354226

Logarithm rules

1 $\log a + \log b = \log ab$

Examples:

1 $\log 5 + \log 4 = \log 20$

2 $\log 10 + \log 2 + \log 5 = \log 100$

Write the following as the log of a single number or expression.

1 $\log 6 + \log 2$ ______

2 $\log 15 + \log 4$ ______

3 $\log 3 + \log 5 + \log 6$ ______

4 $\log 1 + \log 8$ ______

5 $\log 1 + \log 1 + \log 1$ ______

6 $\log p + \log q + \log r$ ______

Write the following in terms of the sums of logs of separate numbers or variables.

7 $\log 6$ ______

8 $\log 9$ ______

9 $\log pq$ ______

10 $\log abc$ ______

2 $\log a - \log b = \log \frac{a}{b}$

Examples:

1 $\log 20 - \log 4 = \log \frac{20}{4}$

$= \log 5$

2 $\log 8 + \log 3 - \log 2 = \log \frac{8 \times 3}{2}$

$= \log 12$

Write the following as the log of a single number or expression.

1 $\log 12 - \log 2$ ______

2 $\log 15 - \log 3$ ______

3 $\log 6 - \log 1$ ______

4 $\log 2 + \log 6 - \log 3$ ______

5 $\log b - \log c$ ______

6 $\log f + \log g - \log h$ ______

Write the following in terms of differences and sums of logs of separate numbers or variables.

7 $\log \frac{15}{3}$ ______

8 $\log \frac{20}{1}$ ______

9 $\log \frac{7 \times 4}{2}$ ______

10 $\log \frac{n}{p \times q}$ ______

ISBN: 9780170354226

3 $n\log a = \log a^n$

You are given this on the formula sheet.

Examples:

1 $2\log 12 = \log 12^2$
$= \log 144$

2 $\frac{1}{2}\log 9 = \log 9^{\frac{1}{2}}$
$= \log 3$

Write the following as the log of a single number.

1 $2\log 5$

2 $3\log 2$

3 $6\log 10$

4 $\frac{1}{2}\log 49$

5 $\frac{1}{3}\log 125$

6 $\frac{1}{a}\log b$

Putting logarithm rules together

1 It is often useful to rewrite the expression using the logarithm of the smallest number possible.
2 Use the logarithm laws to simplify the expression.

Examples:

1 $\frac{1}{2}\log 64 - 3\log 2 = \log\sqrt{64} - \log 2^3$
$= \log 8 - \log 8$
$= 0$

2 $\dfrac{5\log 2 - \frac{1}{2}\log 16}{\log 32} = \dfrac{\log 32 - \log 4}{5\log 2}$
$= \dfrac{\log 8}{5\log 2}$
$= \dfrac{3\log 2}{5\log 2}$
$= \dfrac{3}{5}$

Simplify the following.

1 $\dfrac{\log 16}{\log 2}$

2 $\log 32 + \log 8$

3 $\log 49 - \log 7$

4 $\dfrac{\log 81}{\log 9}$

ISBN: 9780170354226

5 $\log 1000 - \frac{1}{2}\log 100$

6 $\frac{3\log 4}{\log 16}$

7 $\log 125 + \log 5 - \log 25$

8 $\frac{1}{2}\log 64 - \log 4$

9 $\frac{3\log 4 + \log 16}{2\log 4}$

10 $3\log 2 + \log\left(\frac{1}{8}\right)$

11 $\frac{3\log 2 - \log 4}{\log 16}$

12 $\frac{\log 1000 - \log 100}{2\log 10}$

13 $\log\left(\frac{1}{100}\right) + 3\log 10$

14 $\frac{1}{2}\log\left(\frac{1}{16}\right) + \frac{1}{3}\log 64$

Solving logarithm (exponential) equations

To solve these, rearrange the equation into exponent form.

1 Basic

$\log_b y = x$ so rewrite this as $y = b^x$

Need to find y

Use your calculator to find y

Examples:

1 $\log_2 y = 5$
$y = 2^5$
$y = 32$

2 $\log_7 y = 0$
$y = 7^0$
$y = 1$

3 $\log_{16} y = \frac{1}{2}$
$y = 16^{\frac{1}{2}}$
$y = 4$

4 $\log_{10} y = -2$
$y = 10^{-2}$
$y = 0.01$

Solve the following.

1 $\log_4 y = 2$

2 $\log_2 y = 7$

3 $\log_5 y = 1$

4 $\log_{10} y = 0$

5 $\log_{25} a = \frac{1}{2}$

6 $\log_2 z = -3$

7 $\log_3 x = -2$

8 $\log_{25} x = -\frac{1}{2}$

ISBN: 9780170354226

2 Finding the base

$\log_b y = x$ so rewrite this as $y = b^x$

Need to find b

You will need to figure out the value of b, but check it on your calculator

Examples:

1 $\log_b 64 = 3$

$64 = b^3$

$b = 4$

2 $\log_b 5 = 1$

$5 = b^1$

$b = 5$

3 $\log_b 7 = \frac{1}{2}$

$7 = b^{\frac{1}{2}}$

$7 = \sqrt{b}$

$b = 49$

4 $\log_b 0.01 = -2$

$0.01 = b^{-2}$

$b = 10$

Solve the following.

1 $\log_b 100 = 2$

2 $\log_b 81 = 4$

3 $\log_b 23 = 1$

4 $\log_b 4 = \frac{1}{2}$

5 $\log_m 36 = -2$

6 $\log_x 1 = 0$

7 $\log_x 5 = \frac{1}{3}$

8 $\log_x \frac{1}{10} = -\frac{1}{2}$

ISBN: 9780170354226

3 Finding the exponent

$\log_b y = x$ so rewrite this as $y = b^x$

Need to find x

You will need to figure out the value of x, but check it on your calculator

Examples:

1 $\log_2 64 = x$
$64 = 2^x$
$x = 6$

2 $\log_5 1 = x$
$1 = 5^x$
$x = 0$

3 $\log_{36} 6 = x$
$6 = 36^x$
$x = \frac{1}{2}$

4 $\log_{10} 0.001 = x$
$\frac{1}{1000} = 10^x$
$x = -3$

Solve the following.

1 $\log_2 16 = x$

2 $\log_{10} 10\,000 = x$

3 $\log_6 1 = x$

4 $\log_5 125 = x$

5 $\log_9 9 = z$

6 $\log_4 2 = g$

7 $\log_{16} 4 = f$

8 $\log_{64} \frac{1}{2} = h$

NB You can solve these only if you have 'nice' numbers.
The next section shows you how to solve equations where the numbers are not 'nice'.

ISBN: 9780170354226

4 Finding the exponent when you don't have 'nice' numbers

Use this method when you need to calculate the value of an exponent, and where you don't have 'nice' numbers.

Example: $10^x = 2$

Step 1: Take the log of both sides: $\log 10^x = \log 2$

Step 2: Use log rules to rearrange: $x\log 10 = \log 2$

Step 3: Divide both sides by log 10: $x = \frac{\log 2}{\log 10}$

Step 4: Use your calculator to work out the answer. $x = 0.3010$

Do this on your calculator

Examples:

1

$$5^{x+1} = 28$$
$$\log (5^{x+1}) = \log 28$$
$$(x + 1) \log 5 = \log 28$$
$$x + 1 = \frac{\log 28}{\log 5}$$
$$x + 1 = 2.0704$$
$$x = 1.0704$$

2

$$10^{2n} = 94$$
$$\log 10^{2n} = \log 94$$
$$2n \log 10 = \log 94$$
$$2n = \frac{\log 94}{\log 10}$$
$$2n = 1.9731$$
$$n = 0.9866$$

Solve the following.

1 $4^x = 21$

2 $2^x = 100$

3 $100^x = 2$

4 $0.3^x = 3$

ISBN: 9780170354226

5 $5^x = 15$

6 $10^x = 500$

7 $6^{2x} = 10$

8 $7^{-x} = 29$

9 $3^{2x+1} = 6$

10 $100^{3x} = 8$

11 $8^{3x-1} = 20$

12 $3^x4^x = 12$ (Hint: $3^x4^x = 12^x$)

13 $2^{x+4} = 111$

14 $6^{2x+3} = 1$

ISBN: 9780170354226

Applications of exponential equations

Hints:

- These take the form $A = Pr^n$ where:
  - P = the starting value
  - r = the rate of change
  - n = the number of time periods (often years) over which change occurs
  - A = the final amount
- Rates of change (r):

No change	→	$r = 1$				
Increase	→	$r > 1$	e.g.	An increase of 15%	→	r = 1.15
				An increase of 1%	→	r = 1.01
Decrease	→	$r < 1$	e.g.	A decrease of 20%	→	r = 0.80
				A decrease of 3%	→	r = 0.97

- If you are asked to calculate n, you will need to take the log of both sides.

Example one:
Amanda borrows \$100 000 from her parents so she can buy a house. They increase the amount she owes them by 4% at the end of that year, and each subsequent year for as long as she owns the house. Amanda does not repay any of the initial amount that she borrowed.

a Write an expression for the amount that Amanda owes when she has had the money for n years.

$$\text{Amount owed} = \$100\,000 \times (1.04)^n$$

b How much will she owe them at the end of 5 years?

$$\begin{aligned}\text{Amount owed} &= \$100\,000 \times 1.04^5 \\ &= \$121\,665.29\end{aligned}$$

c Use your expression to find the maximum number of years for which Amanda can borrow the money if she is to keep her debt to her parents at less than \$150 000.

$$\begin{aligned}100\,000 \times 1.04^n &= 150\,000 \\ 1.04^n &= \frac{150\,000}{100\,000} = 1.5 \\ n\log 1.04 &= \log 1.5 \\ n &= \frac{\log 1.5}{\log 1.04} = 10.3380 \text{ years}\end{aligned}$$

ISBN: 9780170354226

Example two:
The area of ice on a large lake reduces after the thaw begins in the spring. The area of ice can be modelled by the equation $A = 2.7 \times (0.92)^t$, where the area is measured in hectares and t is the number of days since the thaw began.

a What area of the lake was covered by ice at the start of the thaw?

'at the start of the thaw' ⟶ $t = 0$

so $A = 2.7 \times (0.92)^0$

$A = 2.7$ hectares

Remember $(\text{anything})^0 = 1$

b What area of ice melts during the first week of the thaw?

$$\begin{aligned}\text{Area melted} &= 2.7 - 2.7 \times (0.92)^7 \\ &= 2.7 - 1.5062 \\ &= 1.1938 \text{ hectares}\end{aligned}$$

c How long will it take for half the ice to melt?

$$\begin{aligned}2.7 \times (0.92)^t &= 1.35 \\ (0.92)^t &= \frac{1.35}{2.7} = 0.5 \\ t\log 0.92 &= \log 0.5 \\ t &= \frac{\log 0.5}{\log 0.92} = 8.3130 \text{ days}\end{aligned}$$

Try these yourself.

1 Matiu borrows \$500 from his parents so that he can buy a phone. His parents increase the amount Matiu owes by 6% at the end of that year, and each subsequent year. Matiu does not repay any of the initial amount that he borrowed.

a Write an expression for the amount that Matiu owes when he has had the money for n years.

b How much will he owe his parents at the end of 3 years?

c If he wants to keep his debt to his parents at less than \$700, how long can he borrow the money for?

ISBN: 9780170354226

2 John bought a car for $3500. It depreciates at 15% per year.

a Write an expression for the value of John's car after *n* years.

b According to your expression, what will it be worth after 5 years?

c If he wants to sell the car before it reaches a value of $2000, what is the maximum time he can keep it?

3 Kiri and her brother Jack inherit $5000 each from their grandmother. Kiri invests hers in a scheme that pays 5% interest at the end of every year. Jack invests his in a scheme that pays 3% interest, which is paid at the end of every six months. In both cases, the interest is added to the initial investment.

a Write expressions for the values of Kiri's and Jack's inheritances after *n* years.

b What will each of their inheritances be worth after 5 years?

c How long will it take for each of their investments to double?

d Write a general expression for the length of time it will take for an investment to double if interest of *x* percent is paid annually.

ISBN: 9780170354226

4 Michaela is not happy about a two square metre patch of moss growing in her lawn. She has read that left untreated, it will increase at 24% per year.

a Write an expression for the area of moss in her lawn after n years, given that it is not treated.

b If the moss is not treated, and the area of moss in her lawn increases at the predicted rate of 24% each year, estimate how much of her lawn will be affected after 3 years.

c The smallest quantity of moss killer that she can buy treats 10 m^2 of lawn. Assuming this rate of growth continues, after how many years will the area of lawn affected by moss reach 10 m^2?

5 A bacterial population starts at 500 000 and increases at 15% per day.

a Write an expression for the number of bacteria after n days.

b If the population continues to grow at this rate, how many more bacteria will there be at the end of 4 days?

c Explain what the following calculation would tell you, with respect to the growing bacterial population: $A = 500\,000(1.15^n - 1)$.

d Assuming this rate of growth continues, after how many days will the population reach 1 200 000?

ISBN: 9780170354226

6 There are 50 000 bees in a hive. The hive is infected by a mite that will reduce the population by 18% per week.

a Write an expression for the number of bees in the hive after n weeks.

b If the population continues to decline at this rate, how many fewer bees will there be at the end of 6 weeks?

c The hive becomes less viable when the number of bees reaches 5000. Assuming that this rate of decline continues, after how many weeks will the number of bees reach 5000?

7 Malaki was given $1000 when he was born. This was invested at an interest rate of 3.2% per year, which is added to his investment at the end of every year. The value of his investment after n years can be modelled by $A = 1000 \times (1.032)^n$, where A is the value of the investment.

a How much will the investment be worth by the time he is 10 years old?

b How much more will the investment be worth by the time he is 20 years old, compared with when he was 10?

c Malaki is calculating $1000 \times 1.032^p(1.032^q - 1)$. With reference to the investment, explain what Malaki is calculating.

d Assuming this rate of growth continues, after how many years will the investment have doubled to $2000?

ISBN: 9780170354226

8 A swimming pool, which normally contains 35 000 L of water, has a small leak and it loses 2% of its capacity per day.

a Write an expression for the volume of water remaining after n days.

b If the water leaks at a constant rate, how much will have leaked from the pool after 12 days?

c After how many days will half the water have leaked out?

9 Mobile phones depreciate by 35% per year. Andrew bought a phone for $1050.

a Write an expression for the value of the phone after n years.

b How much will the phone be worth after he has owned it for 3 years?

c How much less will the phone be worth after he has owned it for 5 years, compared with its value after 3 years?

d Andrew is calculating $1050 \times 0.65^p(1 - 0.65^q)$. With reference to the value of his phone, explain what Andrew is calculating.

e Assuming this rate of depreciation continues, after how many years will the phone be worth $50?

ISBN: 9780170354226

Summary of exponent and logarithm rules

Exponents		Logarithms
Rule:	$b^0 = 1$	$\log_b 1 = 0$
Example:	$10^0 = 1$	$\log_{10} 1 = 0$
Rule:	$b^1 = b$	$\log_b b = 1$
Example:	$10^1 = 10$	$\log_{10} 10 = 1$
Rule:	$b^m \times b^n = b^{m+n}$	$\log_b(mn) = \log_b m + \log_b n$
Example:	$10^3 \times 10^2 = 10^5$	$\log_{10}(10^3 \times 10^2) = \log_{10} 10^5$ $= 5$ $\log_{10}(10^3) + \log_{10}(10^2) = 3 + 2$ $= 5$
Rule:	$\frac{b^m}{b^n} = b^{m-n}$	$\log_b \left(\frac{m}{n}\right) = \log_b m - \log_b n$
Example:	$\frac{10^6}{10^2} = 10^4$	$\log_{10} \left(\frac{10^6}{10^2}\right) = \log_{10} (10^4)$ $= 4$ $\log_{10}(10^6) - \log_{10}(10^2) = 6 - 2$ $= 4$
Rule:	$(b^m)^n = b^{m \times n}$	$\log_b m^n = n\log_b m$
Example:	$(10^2)^3 = 10^6$	$\log_{10}(10^2)^3 = \log_{10} 10^6$ $= 6$ $3\log_{10}(10^2) = 3 \times 2$ $= 6$
Rule:	$b^{-m} = \frac{1}{b^m}$	$\log_b \left(\frac{1}{x}\right) = -\log_b x$
Example:	$10^{-2} = \frac{1}{10^2}$	$\log_{10} \left(\frac{1}{10^2}\right) = \log_{10} (10^{-2})$ $= -2$ $-\log_{10} 100 = -\log_{10} 10^2$ $= -2$

ISBN: 9780170354226

Fractions

Multiplying and dividing fractions

Remember:

- Multiplying fractions $\frac{a}{b} \times \frac{c}{d} = \frac{a \times c}{b \times d} = \frac{ac}{bd}$
- Dividing fractions $\frac{a}{b} \div \frac{c}{d} = \frac{a}{b} \times \frac{d}{c}$

$= \frac{ad}{bc}$

Change the '÷' to a 'x'

And invert the second fraction

Examples:

1 $\frac{x}{8} \times \frac{2}{y} = \frac{2x}{8y} = \frac{x}{4y}$

2 $5 \times \frac{x}{6} = \frac{5x}{6}$

3 $\frac{4a}{3a} \times \frac{3b}{5c} = \frac{12ab}{15ac}$
$= \frac{4b}{5c}$

4 $\frac{x}{8} \div \frac{2}{y} = \frac{x}{8} \times \frac{y}{2}$
$= \frac{xy}{16}$

5 $\frac{2x}{3y} \div \frac{3x}{y} = \frac{2x}{3y} \times \frac{y}{3x}$
$= \frac{2xy}{9xy} = \frac{2}{9}$

6 $\frac{x^2}{5} \times \frac{1}{y^3} \div \frac{x^3}{y^2} = \frac{x^2}{5} \times \frac{1}{y^3} \times \frac{y^2}{x^3}$
$= \frac{1}{5xy}$

Simplify these.

1 $\frac{2x}{5} \times \frac{4}{x}$

____

____

2 $\frac{6xy}{5} \times \frac{2y}{3x}$

____

____

3 $\frac{1}{5ab} \times \frac{2a}{5b}$

____

____

4 $4 \times \frac{3xy}{2y}$

____

____

  ISBN: 9780170354226

5 $\frac{4x^4}{x^2} \times \frac{2}{x}$

6 $\frac{x^3yz^2}{xy^2} \times \frac{yz^3}{x^2y^2}$

7 $\frac{x}{4} \div \frac{y}{3}$

8 $\frac{x^4}{4} \div \frac{x^3}{12}$

9 $\frac{4b}{2a} \div \frac{3b}{5}$

10 $x \div \frac{4}{y}$

11 $\frac{(2x)^3}{x^3} \div \frac{3x^2}{2}$

12 $\frac{(3xy)^2 \times (4y)^2}{xy^3} \div 4x^2$

13 $\frac{(6x)^2y}{y^2} \div \frac{3y}{2x}$

14 $\frac{3}{x} \times \frac{y}{2} \div 3xy$

15 $\frac{x}{5} \times \frac{1}{3y} \div \frac{4x}{3}$

16 $\frac{xy}{4} \times \frac{7}{2x} \times \frac{y}{7x}$

17 $\frac{4x^2y}{5y} \times \frac{y}{4x} \div \frac{3y}{5x}$

18 $\frac{(5x^2y)^3}{(3y)^2} \times \frac{(3x^2y)^2}{(4x^3y^4)^3}$

Adding and subtracting fractions

Remember:

1 Adding or subtracting fractions with the **same denominator** is straightforward:

$$\frac{a}{b} + \frac{c}{b} = \frac{a+c}{b} \quad \text{or} \quad \frac{a}{b} - \frac{c}{b} = \frac{a-c}{b}$$

2 Where fractions have **different denominators**, multiply both the numerator and the denominator by a number (or variable) that will create equal denominators.

$$\frac{a}{b} + \frac{c}{d} = \frac{a}{b} \times \frac{d}{d} + \frac{c}{d} \times \frac{b}{b}$$

$\frac{d}{d} = 1$ and $\frac{b}{b} = 1$

$$= \frac{ad}{bd} + \frac{bc}{bd}$$

$$= \frac{ad+bc}{bd}$$

Examples:

1

$$\frac{x}{8} + \frac{x}{3} = \frac{3x}{24} + \frac{8x}{24}$$

$$= \frac{11x}{24}$$

2

$$\frac{x}{2} + x = \frac{x}{2} + \frac{2x}{2}$$

$$= \frac{3x}{2}$$

3

$$\frac{2}{x} - \frac{1}{y} = \frac{2y}{xy} - \frac{1x}{xy}$$

$$= \frac{2y-x}{xy}$$

4

$$\frac{2x+3}{2x} - \frac{1x}{3x} = \frac{6x+9}{6x} - \frac{2x}{6x}$$

$$= \frac{6x+9-2x}{6x}$$

$$= \frac{4x+9}{6x}$$

5

$$\frac{2}{x+3} + \frac{1}{x-2} = \frac{2(x-2)}{(x+3)(x-2)} + \frac{1(x+3)}{(x-2)(x+3)}$$

$$= \frac{2x-2+x+3}{x^2+x-6}$$

$$= \frac{3x+1}{x^2+x-6}$$

Simplify these.

1 $\frac{3}{y} + \frac{4}{y}$

____

____

2 $\frac{2}{x} + \frac{6}{y}$

____

____

ISBN: 9780170354226

3 $\frac{y}{4} - \frac{x}{6}$

4 $\frac{2}{xy} - \frac{x}{5}$

5 $\frac{xy}{3} - \frac{1}{5}$

6 $4 - \frac{x}{6}$

7 $\frac{1}{xy} - 2$

8 $\frac{3}{x^2} + \frac{y}{xy}$

9 $\frac{1}{y^2} + \frac{x}{3}$

10 $\frac{4y}{x} + \frac{3}{x^2y}$

11 $\frac{z}{x} + \frac{x}{y} + \frac{y}{z}$

12 $\frac{1}{x} + \frac{x}{xy} - \frac{2y}{3x}$

13 $\frac{1}{x} + \frac{3}{x-1}$

14 $\frac{2}{x-4} + \frac{5}{x+3}$

15 $\frac{1}{x-3} - \frac{x}{x+2}$

16 $\frac{2(x-1)}{x+4} - \frac{1}{x-3}$

ISBN: 9780170354226

Solving equations involving fractions

These become straightforward if you multiply **every term** on both sides by the lowest multiple of the denominators.

Examples:

1 $\frac{3x}{8} - 2 = \frac{x}{3}$

Step 1: Multiply **every term** by 24 because 3 x 8 = 24: $\frac{3x}{8} \times \frac{24}{1} - 2 \times 24 = \frac{x}{3} \times \frac{24}{1}$

Step 2: Simplify: $9x - 48 = 8x$

$x = 48$

2 $\frac{1}{3x} + 3 = \frac{5}{2x}$

$\frac{1}{3x} \times \frac{6x}{1} + 3 \times 6x = \frac{5}{2x} \times \frac{6x}{1}$

$2 + 18x = 15$

$x = 0.72$

3 $4 + \frac{x-3}{2} = \frac{2x+1}{3}$

$4 \times 6 + \frac{x-3}{2} \times \frac{6}{1} = \frac{2x+1}{3} \times \frac{6}{1}$

$24 + 3(x-3) = 2(2x+1)$

$24 + 3x - 9 = 4x + 2$

$x = 13$

Solve the following equations.

1 $\frac{3x}{5} = \frac{2}{3}$

2 $\frac{7x}{10} = \frac{2}{5}$

3 $\frac{5x}{7} = 2$

4 $\frac{3x}{5} + 1 = \frac{x}{2}$

5 $\frac{2}{x} + 1 = \frac{8}{x}$

6 $\frac{x+3}{4} + \frac{2x-1}{3} = 5$

ISBN: 9780170354226

Solving linear inequations

Inequations have **<**, **≤**, **>** or **≥** signs

You solve an inequation in **exactly** the same way as you solve an equation **except**: if you need to **multiply or divide** the equation by a **negative number**, you must **reverse the sign**.

Examples:

1 $2x < 18$

$x < \frac{18}{2}$

$x < 9$

2 $4x + 1 \geq 17$

$4x \geq 17 - 1$

$4x \geq 16$

$x \geq \frac{16}{4}$

$x \geq 4$

3 $\frac{3x}{2} > 4(x + 3)$

$\frac{3x}{2} > 4x + 12$

$3x > 8x + 24$

$-5x > 24$

$x < \frac{24}{-5}$

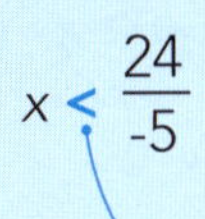

Dividing by a negative number ⟶ **reverse** the sign

4 $\frac{y}{8} - \frac{y}{3} < \frac{5}{12}$

$\frac{y}{8} \times \frac{24}{1} - \frac{y}{3} \times \frac{24}{1} < \frac{5}{12} \times \frac{24}{1}$

$3y - 8y < 10$

$-5y < 10$

$\frac{-5y}{-5} > \frac{10}{-5}$

$y > -2$

Solve the following equations.

1 $3x \geq 21$

____

____

2 $2x - 5 < 21$

____

____

3 $2x + 1 \geq 12 + x$

____

____

____

4 $\frac{x - 1}{2} \geq 5x + 3$

____

____

____

5 $2(x + 4) < 3(4 - 5x)$

____

____

____

6 $-1(8 - 3x) \geq 9(2x - 5) + 7$

____

____

____

ISBN: 9780170354226

7 $\frac{3x}{2} + \frac{2x}{3} < 10$

8 $4 + \frac{x + 3}{3} \geq x - 1$

9 $2(5 - x) > \frac{4x + 7}{3} + 1$

10 $\frac{x - 2}{2} < \frac{3}{4}(4x - 5) - 2$

11 $4 + \frac{6x}{5} > 7 - \frac{5 + 2x}{2}$

12 $\frac{13}{24} - 2x \geq \frac{2 - 7x}{4}$

13 $\frac{7(2x + 1)}{5} > \frac{5(2x - 1)}{3} + \frac{4 - x}{2}$

14 $\frac{8(x - 3)}{5} - \frac{7(4 + 3x)}{2} > \frac{-1(6 + x)}{4}$

ISBN: 9780170354226

Polynomials

Expanding

1 Two brackets

Remember **FOIL:**

Firsts	$(x + 3)(2x - 5)$ ⟶	$2x^2$
Outers	$(x + 3)(2x - 5)$ ⟶	$-5x$
Inners	$(x + 3)(2x - 5)$ ⟶	$+6x$
Lasts	$(x + 3)(2x - 5)$ ⟶	-15

So $(x + 3)(2x - 5) = 2x^2 - 5x + 6x - 15$
$= 2x^2 + x - 15$

Your teacher may show you other methods for this.

Examples:

1 $(2x - 4)(3x - 5) = 6x^2 - 10x - 12x + 20$
$= 6x^2 + 22x + 20$

2 $(2x + 3)(2x - 3) = 4x^2 - 6x + 6x - 9$
$= 4x^2 + 0x - 9$
$= 4x^2 - 9$

3 $(x + 7)^2 = (x + 7)(x + 7)$
$= x^2 + 7x + 7x + 49$
$= x^2 + 14x + 49$

4 $(4x - 2)^2 = (4x - 2)(4x - 2)$
$= 16x^2 - 8x - 8x + 4$
$= 16x^2 - 16x + 4$

Expand and simplify these.

1 $(x - 3)(x + 4)$

____

____

2 $(x - 7)(x - 2)$

____

____

3 $(5 + x)(x + 8)$

____

____

4 $(x + 8)^2$

____

____

ISBN: 9780170354226

5 $(2x + 4)(x - 2)$

6 $(3x - 3)(2x + 6)$

7 $(3x - 5)^2$

8 $(3 - x)^2$

9 $(4 + 2x)(2 - 6x)$

10 $(x^2 + 3)(x - 6)$

11 $(5x - 2)(4x + 3)$

12 $(7 - 2x)(4 + 3x)$

13 $(x + 1)(x - 1)$

14 $(5 - 2x)(5 + 2x)$

15 $(4x - 3)(4x + 3)$

16 $(7x + 2)^2$

 ISBN: 9780170354226

2 Three brackets

Steps:

1 If necessary, write as three sets of brackets. $(x + 3)(2x - 5)^2 = (x + 3)(2x - 5)(2x - 5)$

2 Multiply any two of the sets of brackets using **FOIL**. $= (x + 3)(4x^2 - 10x - 10x + 25)$

3 Collect the like terms. $= (x + 3)(4x^2 - 20x + 25)$

4 Multiply the result by each term in the third set of brackets. $= 4x^3 - 20x^2 + 25x + 12x^2 - 60x + 75$

5 Collect the like terms. $= 4x^3 - 8x^2 - 35x + 75$

Examples:

1 $(2x + 3)(3x - 5)(2x + 4) = (6x^2 - 10x + 9x - 15)(2x + 4)$

$= (6x^2 - 1x - 15)(2x + 4)$

$= 12x^3 + 24x^2 - 2x^2 - 4x - 30x - 60$

$= 12x^3 + 22x^2 - 34x - 60$

2 $(3x - 2)^3 = (3x - 2)(3x - 2)(3x - 2)$

$= (9x^2 - 6x - 6x + 4)(3x - 2)$

$= (9x^2 - 12x + 4)(3x - 2)$

$= 27x^3 - 18x^2 - 36x^2 + 24x + 12x - 8$

$= 27x^3 - 54x^2 + 36x - 8$

Expand and simplify these.

1 $(x + 1)(x + 3)(x + 2)$

2 $(x + 4)(x - 6)(x + 2)$

3 $(3 - x)(x - 2)(x + 2)$

4 $(x + 3)^3$

ISBN: 9780170354226

5 $(x - 5)^3$

6 $(2x - 4)^3$

7 $(x + 4)^2(x - 2)$

8 $(3x - 2)(2x - 1)(2x + 3)$

9 $(2x + 3)(x + 5)^2$

10 $(x - 1)(2x - 3)^2$

11 $x(x - 6)(x + 2)$

12 $2x(x - 4)(x - 3)$

13 $(x - 5)(x^2 + x - 2)$

14 $(3x^2 + 2x - 2)(2x - 3)$

ISBN: 9780170354226

Factorising quadratics

1 Where the coefficient of x^2 is 1

Steps:

1 List all the factors of the constant: $x^2 + 3x - 40$

1, 40
2, 20
4, 10
5, 8

2 Select the pair that could add or subtract to give the coefficient of x: $x^2 + 3x - 40$

$-5 + 8 = +3$

3 The factors are $(x - 5)(x + 8)$

4 **Check** your answer by expanding the brackets using **FOIL** — you should get the original expression: $(x - 5)(x + 8) = x^2 + 3x - 40$

Examples:

1 $x^2 - 11x + 30 = (x - 5)(x - 6)$

1, 30
2, 15
3, 10
-5, -6

2 $x^2 - 7x - 18 = (x + 2)(x - 9)$

1, 18
+2, -9
3, 6

3 $x^2 - 25 = x^2 + 0x - 25 = (x + 5)(x - 5)$

1, 25
+5, -5

Insert a 'fake' x term

Factorise the following.

1 $x^2 + 6x + 8$

____

____

2 $x^2 + 13x + 36$

____

____

3 $x^2 + 8x + 7$

____

____

4 $x^2 + 4x - 12$

____

____

ISBN: 9780170354226

5 $x^2 + 17x + 52$

6 $x^2 - 9x + 18$

7 $x^2 - 2x - 24$

8 $24 + 11x + x^2$

9 $x^2 - 64$

10 $1 - x^2$

11 $x^2 - 10x + 25$

12 $x^2 - 25$

13 $9 - 16y^2$

14 $-x^2 + 49$

15 $-x^2 - 5x + 6$

16 $-x^2 + 5x - 6$

17 $-x^2 + 6x - 8$

18 $25 - 9x^2$

ISBN: 9780170354226

2 Where the coefficient of x^2 is not 1, but there is a common factor

Steps:

1 Take out the common factor first: $2x^2 - 4x - 6 = 2(x^2 - 2x - 3)$

2 Factorise the part inside the bracket: $= 2(x - 3)(x + 1)$

Examples:

1 $6x^2 + 18x - 60 = 6(x^2 + 3x - 10)$
$= 6(x + 5)(x - 2)$

2 $-3x^2 + 18x - 24 = -3(x^2 - 6x + 8)$
$= -3(x - 4)(x - 2)$

3 $-7x^2 + 28 = -7(x^2 - 4)$
$= -7(x + 2)(x - 2)$

4 $27x^2 - 75 = 3(9x^2 - 25)$
$= 3(3x + 5)(3x - 5)$

Factorise the following.

1 $4x^2 + 16x - 48$

____

____

2 $-8x^2 + 8$

____

____

3 $20x + 4x^2 + 16$

____

____

4 $200 - 8x^2$

____

____

5 $-5x^2 - 10x - 5$

____

____

6 $-4x^2 - 16x - 12$

____

____

7 $2x^2 - 2x - 12$

____

____

8 $5x^2 - 180$

____

____

ISBN: 9780170354226

3 Where the coefficient of x^2 is not 1, and there is no common factor

Examples:

1 Factorise $3x^2 + 11x + 10$

This does not have 1 as the coefficient of x^2 and it has no common factor.

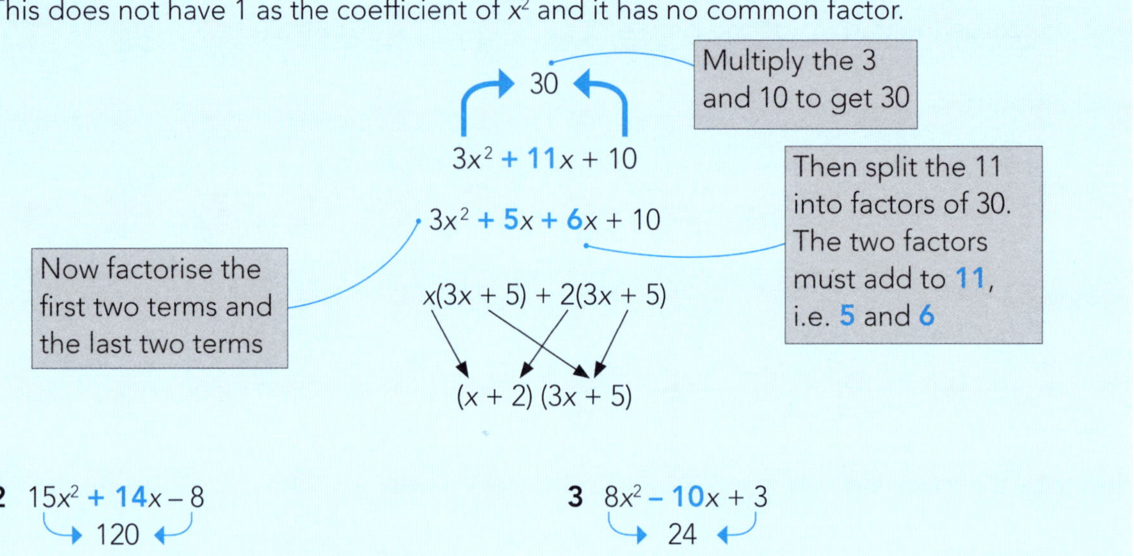

2 $15x^2 + 14x - 8$

$\quad$ 120

$15x^2 + 20x - 6x - 8$

$5x(3x + 4) - 2(3x + 4)$

$(5x - 2)(3x + 4)$

Watch the sign!

3 $8x^2 - 10x + 3$

$\quad$ 24

$8x^2 - 4x - 6x + 3$

$4x(2x - 1) - 3(2x - 1)$

$(4x - 3)(2x - 1)$

Factorise the following.

1 $2x^2 + 11x + 12$

2 $2x^2 + 13x + 6$

3 $4x^2 + 16x + 15$

4 $2x^2 - 5x - 12$

5 $3x^2 - 20x + 12$

6 $4x^2 - 10x + 6$

ISBN: 9780170354226

7 $4x^2 - 15x - 25$

8 $4x^2 - 6x + 2$

9 $10x^2 + 13x - 3$

10 $9x^2 - 9x + 2$

11 $-3x^2 - 4x + 4$

12 $-2x^2 - 2x + 4$

13 $-2x^2 - 1x + 3$

14 $3x^2 - 25x - 18$

15 $12x^2 + 5x - 2$

16 $2x^2 - 9x - 56$

17 $-2 + 5x^2 - 9x$

18 $-x + 6x^2 - 1$

19 $6 + 13x + 6x^2$

20 $3 + 2x^2 - 12x + 15$

ISBN: 9780170354226

Simplifying rational quadratic expressions

Factorise the top and bottom, then cancel the common factors.

Examples:

1 $\dfrac{x^2 + 7x + 12}{x + 3} = \dfrac{(x + 4)\cancel{(x + 3)}}{\cancel{x + 3}}$

$= x + 4$

2 $\dfrac{9 - x^2}{x^2 - x - 6} = \dfrac{(3 - x)(3 + x)}{(x - 3)(x + 2)}$

$= \dfrac{-\cancel{(x - 3)}(3 + x)}{\cancel{(x - 3)}(x + 2)}$

$= \dfrac{-(3 + x)}{x + 2}$

Simplfy these.

1 $\dfrac{2x + 8}{x + 4}$

2 $\dfrac{x - 1}{4x - 4}$

3 $\dfrac{x - 3}{3 - x}$

4 $\dfrac{2x - 12}{3x - 18}$

5 $\dfrac{x^2 + 10x + 16}{x + 2}$

6 $\dfrac{x^2 - 7x + 12}{x - 3}$

7 $\dfrac{x^2 + 4x + 3}{x^2 + 2x - 3}$

8 $\dfrac{x^2 + 4x + 4}{x^2 + 6x + 8}$

9 $\dfrac{2x^2 - 10x - 12}{x^2 - 8x + 12}$

10 $\dfrac{4x^2 - 48 + 16x}{-10x + 2x^2 + 12}$

ISBN: 9780170354226

Solving quadratic equations

1 By factorising

Examples:

1 $2x^2 + 5x = 3$

Step 1: Rearrange so that 0 is on the right: $2x^2 + 5x - 3 = 0$

Step 2: Factorise: $(2x - 1)(x + 3) = 0$

This means one bracket **multiplied** by the other. If $a \times b = 0$, then either $a = 0$ or $b = 0$.

Step 3: Solve by considering what happens when each bracket equals 0.

Either $(2x - 1) = 0$
$2x = 1$
$x = \frac{1}{2}$

or $(x + 3) = 0$
$x = -3$

So, if $2x^2 + 5x = 3$, then x is either $\frac{1}{2}$ or -3.

- Most quadratic equations that you come across will have two solutions. However, some have just one solution, and some have no real solutions.
- In a practical situation there may be two solutions, but one may not work.

2
$x^2 - 3x - 10 = 0$
$(x + 2)(x - 5) = 0$
$x = -2$ or 5

3
$x^2 + 10x + 25 = 0$
$(x + 5)(x + 5) = 0$
$x = -5$

4
$x^2 - 16 = 0$
$(x + 4)(x - 4) = 0$
$x = -4$ or 4

5
$6x^2 + 5x - 6 = 0$
$(2x + 3)(3x - 2) = 0$
Either $2x + 3 = 0$
$x = -\frac{3}{2}$
or $3x - 2 = 0$
$x = \frac{2}{3}$

Solve the following equations.

1 $x^2 + 5x + 6 = 0$

2 $x^2 - 3x - 10 = 0$

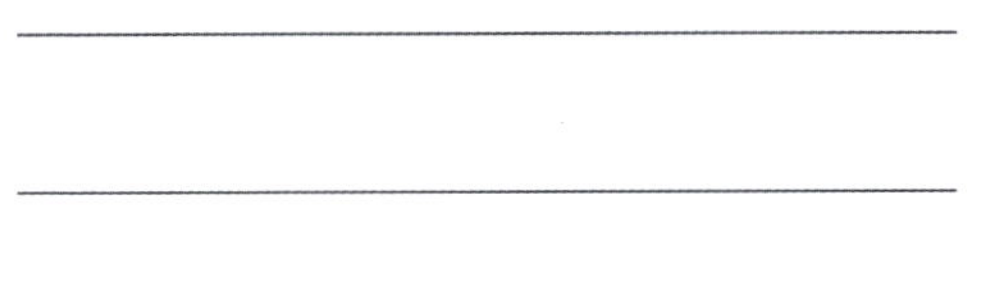

ISBN: 9780170354226

3 $x^2 - 10x + 21 = 0$

4 $x^2 - 9x = 0$

5 $x^2 + 6x + 9 = 0$

6 $x^2 - 36 = 0$

7 $4x^2 - 81 = 0$

8 $2x^2 - 10x - 48 = 0$

9 $2x^2 + 3x - 2 = 0$

10 $15x^2 + 7x - 2 = 0$

11 $21x^2 - x - 2 = 0$

12 $-3x^2 + 2x + 8 = 0$

13 $4 + 9x + 5x^2 = 0$

14 $1 + 7x + 6x^2 = 0$

ISBN: 9780170354226

2 Using the quadratic formula

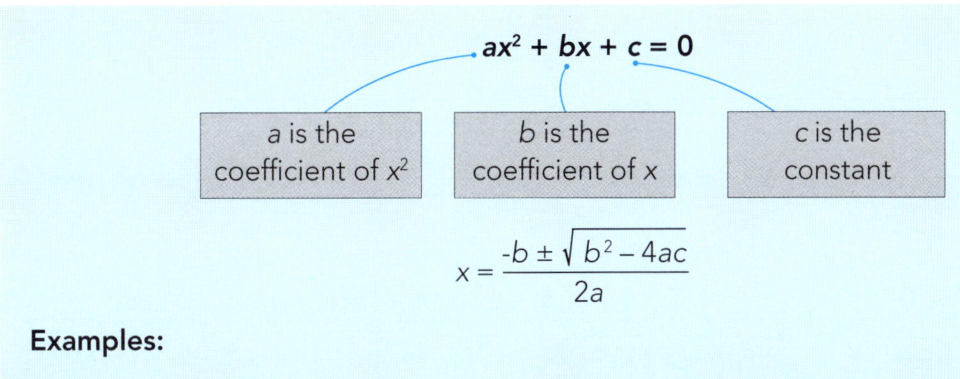

$$x = \frac{-b \pm \sqrt{b^2 - 4ac}}{2a}$$

Examples:

1 $x^2 + 5x + 6 = 0$

$a = 1 \quad b = 5 \quad c = 6$

$$x = \frac{-5 \pm \sqrt{5^2 - 4 \times 1 \times 6}}{2 \times 1}$$

$$x = \frac{-5 \pm \sqrt{1}}{2}$$

$x = -2$ or $x = -3$

2 $2x^2 + 9x + 3 = 0$

$a = 2 \quad b = 9 \quad c = 3$

$$x = \frac{-9 \pm \sqrt{9^2 - 4 \times 2 \times 3}}{2 \times 2}$$

$$x = \frac{-9 \pm \sqrt{57}}{4}$$

$x = -0.3625$ or $x = -4.1375$

3 $4x^2 - 12x + 9 = 0$

$a = 4 \quad b = -12 \quad c = 9$

$$x = \frac{-(-12) \pm \sqrt{(-12)^2 - 4 \times 4 \times 9}}{2 \times 4}$$

$$x = \frac{12 \pm \sqrt{0}}{8}$$

$x = 1.5$

One solution only

4

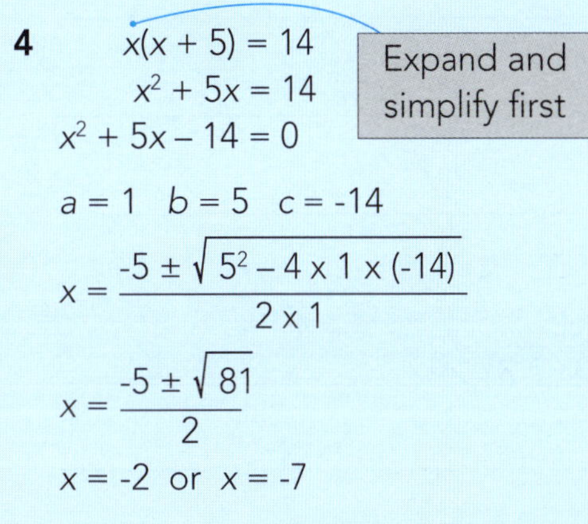

$x(x + 5) = 14$

$x^2 + 5x = 14$

$x^2 + 5x - 14 = 0$

$a = 1 \quad b = 5 \quad c = -14$

$$x = \frac{-5 \pm \sqrt{5^2 - 4 \times 1 \times (-14)}}{2 \times 1}$$

$$x = \frac{-5 \pm \sqrt{81}}{2}$$

$x = -2$ or $x = -7$

Solve these using the quadratic formula.

1 $x^2 + 7x + 12 = 0$

____

____

____

2 $x^2 + 4x + 3 = 0$

____

____

____

ISBN: 9780170354226

3 $x^2 + 6x - 16 = 0$

4 $x^2 - 5x + 4 = 0$

5 $2x^2 + 9x + 4 = 0$

6 $2x^2 - 4x + 1 = 0$

7 $3x^2 - 2x - 6 = 0$

8 $3x^2 + 4x - 6 = 0$

9 $4 - 6x + x^2 = 0$

10 $x^2 - 7x = -4$

11 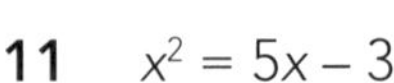 $x^2 = 5x - 3$

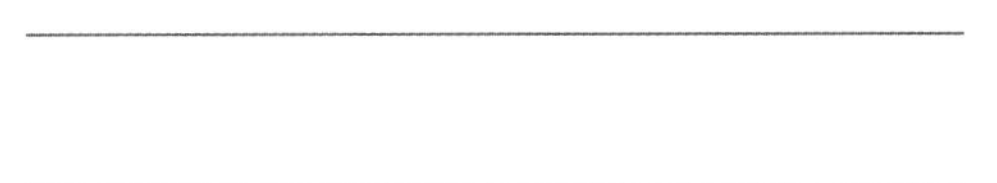

12 $9 = x^2 - 2x$

 ISBN: 9780170354226

3 On a calculator

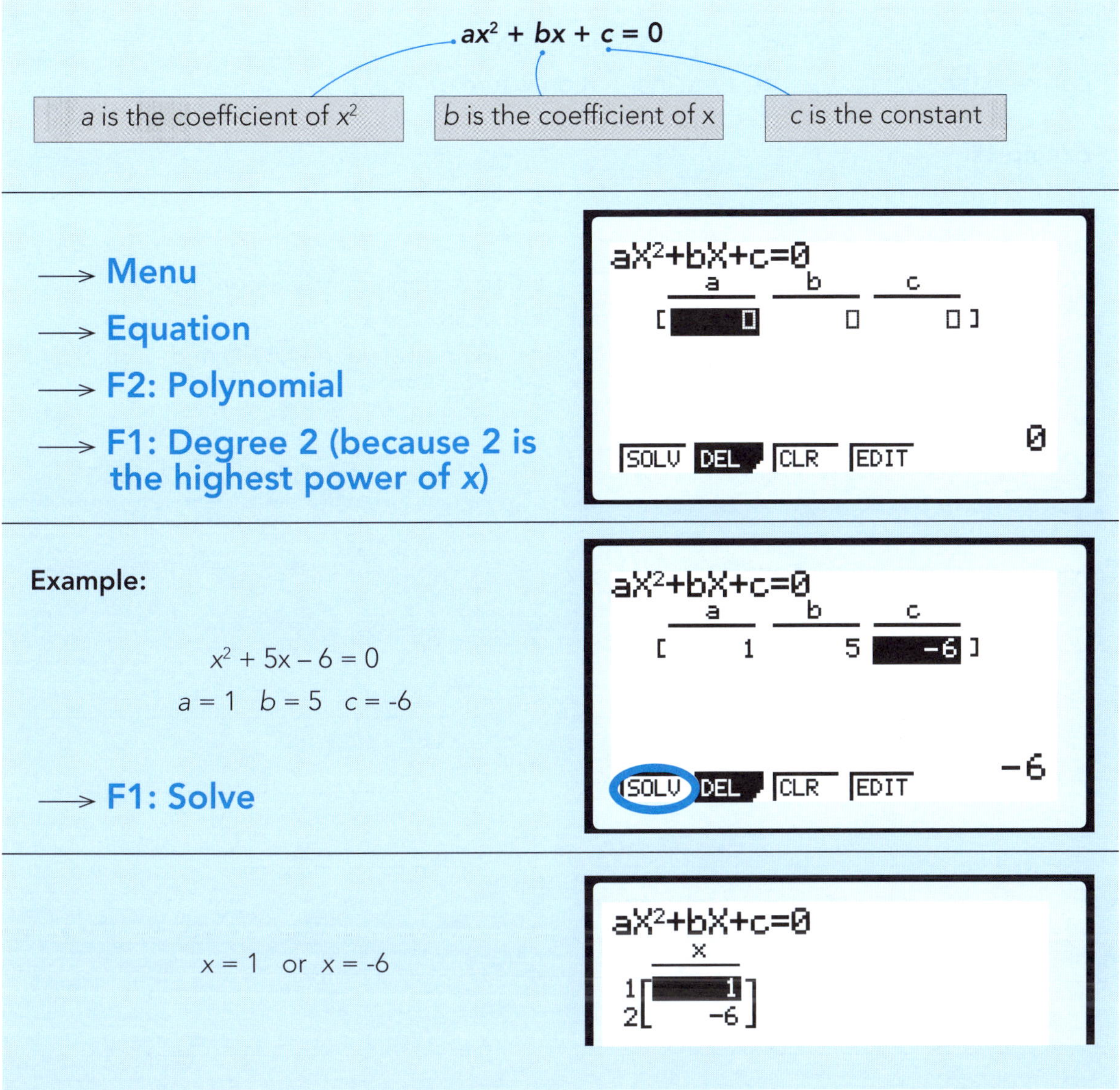

Solve the following using your graphics calculator. Compare your answers with the previous exericse.

1	$x^2 + 7x + 12 = 0$ __________	**2**	$x^2 + 4x + 3 = 0$ __________
3	$x^2 + 6x - 16 = 0$ __________	**4**	$x^2 - 5x + 4 = 0$ __________
5	$2x^2 + 9x + 4 = 0$ __________	**6**	$2x^2 - 4x + 1 = 0$ __________
7	$3x^2 - 2x - 6 = 0$ __________	**8**	$3x^2 + 4x - 6 = 0$ __________
9	$4 - 6x + x^2 = 0$ __________	**10**	$x^2 - 7x = -4$ __________
11	$x^2 = 5x - 3$ __________	**12**	$9 = x^2 - 2x$ __________

ISBN: 9780170354226

Solving quadratic inequations

Solve these as if you are solving a quadratic equation, then substitute a value that is between the solutions to see if the inequation is true or not.

Examples:

1 $x^2 - 3x - 10 < 0$

$(x + 2)(x - 5) < 0$

$x = -2$ or 5

$x = 0$ lies between -2 and 5, and is easy to substitute:

$x = 0 \Rightarrow 0^2 - 3(0) - 10 < 0$

$-10 < 0$ is true.

$\therefore$ **-2 < x < 5**

2 $x^2 + 7x + 11 > 0$

Using a graphics calculator or the formula:

$x = -2.3820$ or $x = -4.6180$

$x = -3$ lies between -4.6180 and -2.3820.

$x = -3 \Rightarrow (-3)^2 + 7(-3) + 11 > 0$

$-1 > 0$ is **not** true.

$\therefore$ **x < -4.6180 or x > -2.3820**

Solve the following inequations.

1 $x^2 - 4x - 12 < 0$

2 $x^2 - 6x + 8 < 0$

3 $x^2 + 9x + 14 \geq 0$

4 $x^2 - 4x - 11 \leq 0$

5 $x^2 - 5x + 3 > 0$

6 $x^2 + 9x + 15 < 0$

ISBN: 9780170354226

Solving rational quadratic equations

- **If possible, factorise and cancel first** — otherwise you will get two solutions, one of which doesn't work.
- Then cross-multiply.
- Solve.

Example:

By factorising and cancelling first:

$$\frac{x^2 - x - 6}{x^2 - 7x + 12} = 7$$
$$\frac{(x-3)(x+2)}{(x-3)(x-4)} = 7$$
$$\frac{x+2}{x-4} = 7$$
$$7x - 28 = x + 2$$
$$6x = 30$$
$$x = 5$$

By cross-multiplying first:

$$\frac{x^2 - x - 6}{x^2 - 7x + 12} = 7$$
$$7x^2 - 49x + 84 = x^2 - x - 6$$
$$6x^2 - 48x + 90 = 0$$
$$x^2 - 8x + 15 = 0$$
$$(x-3)(x-5) = 0$$
$$x = 3 \text{ or } x = 5$$

Note the **extra solution**

Substituting **$x = 3$** into line 2 of the first method, $\frac{(x-3)(x+2)}{(x-3)(x-4)} = 7$, produces a denominator of 0, so **$x = 3$** is **not** a solution.

Solve the following equations.

1 $\frac{x^2 + 4x + 3}{x^2 - 2x - 3} = 5$

2 $\frac{x^2 + 8x + 16}{x^2 - 16} = 3$

3 $\frac{x^2 + x - 12}{x^2 + 6x + 8} = 6$

4 $\frac{x^2 - 3x - 10}{x^2 - 8x + 16} = 2$

5 $\frac{2x^2 - 2x}{x^2 - 5x - 6} = -1$

6 $\frac{x^2 + 3x + 2}{x(x-1)} = 6$

ISBN: 9780170354226

Roots of equations

1 Calculating the value of the discriminant and number of roots

- Quadratic equations have a maximum of two solutions, or points where the parabola crosses the x-axis. These solutions are also known as **roots**.
- We use the term 'real' roots, because at Level 3 you will learn about imaginary or unreal roots.
- You can tell how many roots there are by the value of $b^2 - 4ac$ in the quadratic formula. This is known as the **discriminant** (Δ).

There are three possibilities for the number of real roots:

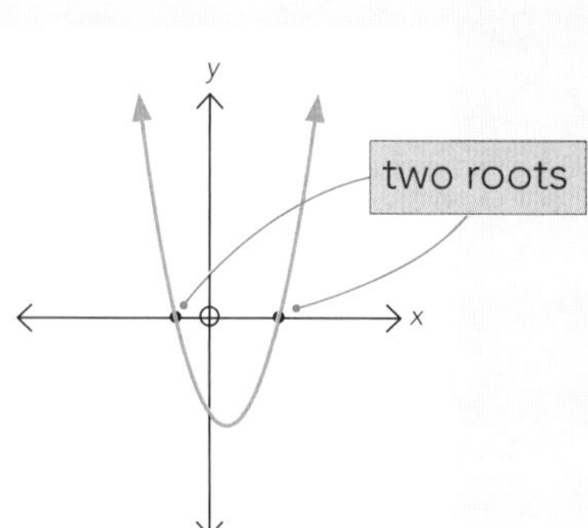

$\sqrt{+ve} \rightarrow$ 2 real roots

$\Delta = b^2 - 4ac$ **> 0**

So $x = \dfrac{-b + \sqrt{\text{positive number}}}{2a}$

or $x = \dfrac{-b - \sqrt{\text{positive number}}}{2a}$

So there are two real distinct roots.

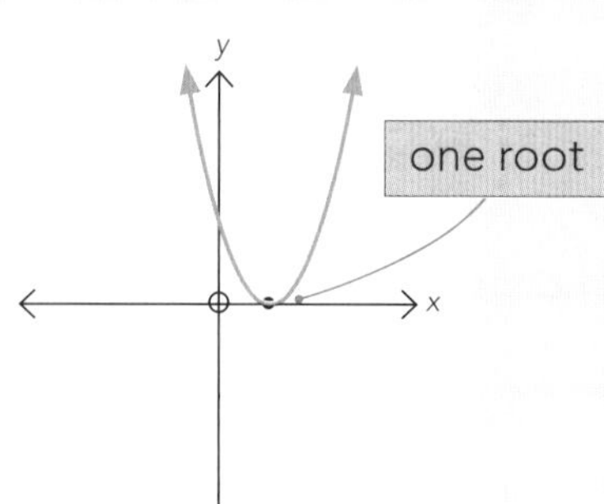

$\sqrt{0} \rightarrow$ 1 real root

$\Delta = b^2 - 4ac$ **= 0**

So $x = \dfrac{-b \pm \sqrt{0}}{2a} = \dfrac{-b}{2a}$

So there is one real root.
(This could be described as two identical roots.)

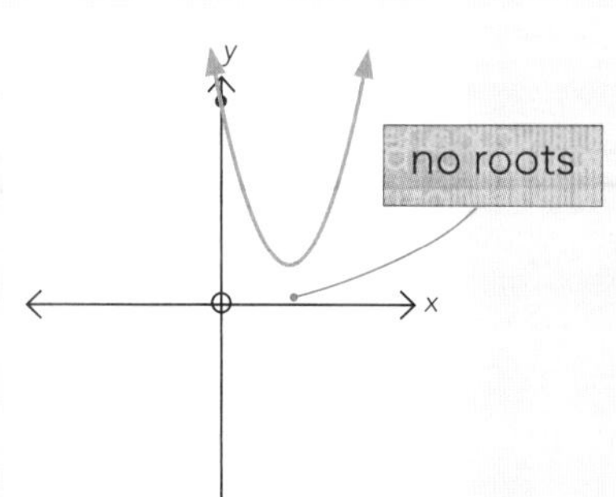

$\sqrt{-ve} \rightarrow$ 0 real roots

$\Delta = b^2 - 4ac$ **< 0**

But $\sqrt{\text{negative number}}$ is not possible

So there are no real roots.

Example: Calculate the value of the discriminant for $10x^2 - x - 3 = 0$, and use it to determine how many roots it has.

$$\begin{aligned}\Delta = b^2 - 4ac &= (-1)^2 - 4 \times 10 \times (-3)\\ &= 1 + 120\\ &= +121\end{aligned}$$

$\therefore$ the discriminant is positive, so there are two real roots.

ISBN: 9780170354226

Calculate the value of the discriminant and state the number of real roots.

1 $2x^2 + 5x - 12 = 0$

2 $9x^2 - 12x + 4 = 0$

3 $2x^2 - 7x + 10 = 0$

4 $4x^2 + 5x + 2 = 0$

5 $6x^2 + 11x + 4 = 0$

6 $x^2 - x + 0.25 = 0$

7 $4x^2 + 20x + 25 = 0$

8 $2x^2 - 3x + 6 = 0$

9 $0.5x^2 - 3x + 1 = 0$

10 $19x^2 + 7 = 0$

11 $7x^2 - 20x = 0$

12 $0.01x^2 - 0.6x + 9 = 0$

ISBN: 9780170354226

Sometimes you might need to rearrange the equation first.

Examples:

1 Calculate the value of the discriminant and state the number of real roots for the equation $5x^2 = 3x - 6$.

Step 1: Reorganise into $ax^2 + bx + c = 0$ form. $5x^2 - 3x + 6 = 0$

Step 2: Calculate the discriminant.

$$b^2 - 4ac = (-3)^2 - 4 \times 5 \times 6 = 9 - 120 = -111$$

$\therefore$ $5x^2 = 3x - 6$ has no real roots

2 Calculate the value of the discriminant and state the number of real roots for the equation $12x = 4x^2 + 9$.

Step 1: Reorganise into $ax^2 + bx + c = 0$ form. $4x^2 - 12x + 9 = 0$

Step 2: Calculate the discriminant.

$$b^2 - 4ac = (-12)^2 - 4 \times 4 \times 9 = 144 - 144 = 0$$

$\therefore$ $12x = 4x^2 + 9$ has one real root

Calculate the value of the discriminant and state the number of real roots.

13 $3x^2 = 20 - 11x$

14 $6x^2 = 2 - 5x$

15 $12x = 9 + 4x^2$

16 $(5x + 1)^2 = -7$

17 $9x^2 - 2 = 6x - 3$

18 $20x^2 + 7 = 3x^2 - 13$

ISBN: 9780170354226

2 Finding values for *a*, *b* or *c* when given the number of roots

Sometimes you might be asked to find the values of variables in the equation.

Don't forget:

- The square root of a number can be positive or negative.
- When you multiply or divide an inequality by a negative number, flip the sign
- The roots occur when $y = 0$.
- Often you will need to convert the equation into the $ax^2 + bx + c$ format.

Examples:

1 For what values of p will the graph of $y = px^2 - 3x + 4$ not cut the x axis?

'not cut the x axis' → no real roots → $b^2 - 4ac$ **< 0**

$b^2 - 4ac = (-3)^2 - 4 \times p \times 4 < 0$

$9 - 16p < 0$

$-16p < -9$

$p > 0.5625$ — Flip sign

So $y = px^2 - 3x + 4$ will not cut the x-axis if $p > 0.5625$

2 The equation $(3x - 1)(x + 5) = k$ has only one real solution. Find the value of k.

Step 1: Convert to $ax^2 + bx + c$ format: $3x^2 + 14x - 5 = k$

$3x^2 + 14x - 5 - k = 0$

$3x^2 + 14x - (5 + k) = 0$ — Watch the sign!

Step 2: One solution → $b^2 - 4ac = 0$: $(14)^2 - (4 \times 3 \times -(5 + k)) = 0$

$196 + 60 + 12k = 0$

$12k = -256$

$k = -21.33$

So $(3x - 1)(x + 5) = k$ has only one solution if $k = -21.3333$

3 For what values of q will the equation $5x^2 + qx + 2 = 0$ have two real roots?

Two real roots → $b^2 - 4ac > 0$

$(q)^2 - 4 \times 5 \times 2 > 0$

$q^2 - 40 > 0$

$q^2 > 40$

$\therefore$ either $q > 6.325$

or $-q > 6.325$

$q < -6.325$

So there will be two real roots if $q > 6.325$ or $q < -6.325$

ISBN: 9780170354226

Try these yourself.

1 For what values of p will the graph of $y = px^2 - 3x + 4$ not cut the x-axis?

2 For what value of q will $2x^2 - 6x + q = 0$ have exactly one real root?

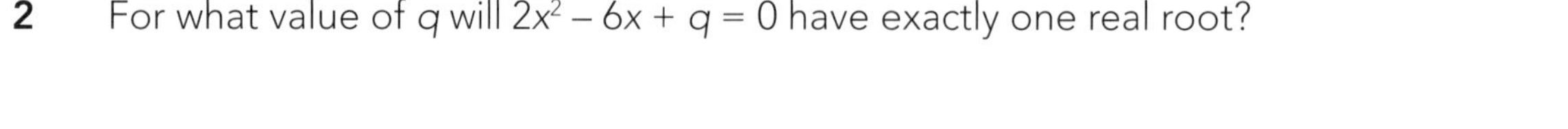

3 For what values of p will $y = px^2 - 5x + 9$ have two real distinct roots?

4 The equation $(3x - 1)(x + 5) = k$ has two identical real solutions. Find the value of k.

5 The equation $(5x + 2)(3x - 1) = k$ has no real solutions. Find the value of k.

 ISBN: 9780170354226

6 Find possible values for q if the equation $8x^2 + 5x + q = 0$ does not cut the x-axis.

7 Find possible values for p if the equation $5x^2 + px + 6 = 0$ has two real distinct roots.

8 The equation $9x + x(2x - 5) + 3 = k$ has no real solutions. Find the values of k.

9 For what values of k will the equation $3x(2x - 5) + 2(1 - x) = k$ have no roots?

10 Find the value of k if the equation $y = x^2 + (k - 3)x + (2 - k)$ has two identical roots.

ISBN: 9780170354226

11 Find the possible values for d if real solutions exist for $3x^2 + (d - 1)x + (d + 4) = 0$.

12 Find possible values for d if no real solutions exist for $x^2 - 3x + 4 - d(x^2 - 1) = 0$.

13 **a** Find expressions, in terms of m and n, for the roots of the equation: $\dfrac{x + m}{x - n} = \dfrac{2m}{x + n}$

b Give an inequality, in terms of m and n, so the equation has two distinct roots.

14 **a** Find expressions, in terms of m and n, for the roots of the equation: $\dfrac{3m - x}{x + n} = \dfrac{2(x + m)}{x - n}$

b Give an inequality, in terms of m and n, so the equation has no distinct roots.

ISBN: 9780170354226

3 Solving when substitution is required

Sometimes you might have to substitute for an expression within an equation.

Hints:

- Remember that $\sqrt{(ax + b)}^2 = (ax + b)$ and $(a^n)^2 = (a^2)^n = a^{2n}$.
- You will usually get two answers — check both because one may not satisfy the original equation.
- Even though you are asked about roots, simply factorising, rather than using the formula, is often the best approach.
- These questions often look much harder than they are.

Examples:

1 The equation $(x + 3) - 2\sqrt{(x + 3)} - 3 = 0$ has only one real solution. Find the value of x. (Hint: Let $p = \sqrt{(x + 3)}$)

$(x + 3) = \sqrt{(x + 3)}^2$ so $p^2 - 2p - 3 = 0$

$(p + 1)(p - 3) = 0$

$\therefore \mathbf{p = -1}$ **or** $\mathbf{p = 3}$

To find the value of x:

1 Let $p = -1$: Then $\sqrt{(x + 3)} = -1$

$x + 3 = (-1)^2 = 1$

$x = -2$

Test this answer: $(-2 + 3) - 2\sqrt{(-2 + 3)} - 3 = -4$

$-4 \neq 0 \therefore p = -1$ and $x = -2$ do not produce a solution.

2 Let $p = 3$: Then $\sqrt{(x + 3)} = 3$

$x + 3 = 9$

$x = 6$

Test this answer: $(6 + 3) - 2\sqrt{6 + 3} - 3 = 9 - 6 - 3 = 0$

The equation balances, so $x = 6$ is the solution.

2 Solve $4^x - (7 \times 2^x) - 8 = 0$ and explain why it has only one real solution. (Hint: Let $p = 2^x$)

$4^x = (2^x)^2$ so $p^2 - 7p - 8 = 0$

$(p + 1)(p - 8) = 0$

$\therefore \mathbf{p = -1}$ **or** $\mathbf{p = 8}$

To find the value of x:

1 Let $p = -1$: Then $2^x = -1$

There is no value for x which could satisfy $2^x = -1 \therefore p = -1$ does not produce a solution.

2 Let $p = 8$: Then $2^x = 8 \rightarrow x = 3$

Test this answer: $4^3 - (7 \times 2^3) - 8 = 64 - 56 - 8 = 0$

The equation balances, so $x = 3$ is the solution.

ISBN: 9780170354226

Try these yourself.

1 The equation $(x + 3) - 2\sqrt{(x + 3)} + 1 = 0$ has only one real solution. Find the value of x and explain why there is only one solution. (Hint: Let $p = \sqrt{(x + 3)}$)

2 The equation $(x - 3) - 2\sqrt{(x - 3)} - 3 = 0$ has two real solutions. Find the values of x. (Hint: Let $p = \sqrt{(x - 3)}$)

3 The equation $(x + 5) - 4\sqrt{(x + 5)} - 12 = 0$ has only one real solution. Find the value of x and explain why there is only one solution.

4 Solve $9^x + (11 \times 3^x) - 12 = 0$ and explain why there is only one real solution. (Hint: Let $p = 3^x$)

ISBN: 9780170354226

5 Solve $16^x + 4^x - 6 = 0$ and explain why there is only one real solution.

6 The equation $4^x - (11 \times 2^x) + 28 = 0$ has two real roots. Calculate the values of these roots. Show your reasoning.

7 By substituting p for x^2 in $x^4 - 2x^2 - 8 = 0$, show that the function has two real distinct roots, and determine their values.

8 Solve $8x^4 - 2x^2 - 3 = 0$. Show your working.

ISBN: 9780170354226

Forming and solving quadratic equations

1 Given information about the equation

Hints:

- Read what the question asks for. Call this x.
- If there are two things to find, call the smaller one x and the bigger one $(x + ?)$.
- Usually you will get two answers. Test both to see if they work. If you have to reject one answer because it doesn't work, explain why.

a Finding unknown numbers

- Even numbers: call these $2x$, e.g. 10, where $x = 5$.
- Odd numbers: call these $2x + 1$, e.g. 15, where $x = 7$.
- Consecutive numbers come after each other: call these x, $x + 1$, $x + 2$, e.g. 6, 7, 8, where $x = 6$.
- Consecutive even numbers: call these $2x$, $2x + 2$, $2x + 4$, e.g. 6, 8, 10, where $x = 3$.
- Consecutive odd numbers: call these $2x + 1$, $2x + 3$, $2x + 5$, e.g. 9, 11, 13, where $x = 4$.
- Age questions: give the youngest the age x.

Example: Matiu thinks of a positive number. He adds 4, squares the result, subtracts 13 times the number he first thought of. His answer is 52. What number was he thinking of?

Call the number he first thought of x.

Adds four,	$x + 4$
squares the result,	$(x + 4)^2$
subtracts 13 times the number he first thought of,	$(x + 4)^2 - 13x$

$$(x + 4)^2 - 13x = 52$$
$$x^2 + 8x + 16 - 13x = 52$$
$$x^2 - 5x - 36 = 0$$
$$(x + 4)(x - 9) = 0$$
$$\therefore x = -4 \text{ or } x = 9$$

His number had to be positive, so -4 is not a possibility, therefore he thought of 9.

Try for yourself.

Mariama thinks of a positive number. She squares it, then multiplies her answer by 5, then subtracts 20 times the number she first thought of. Her answer is 60. What number was she thinking of?

 ISBN: 9780170354226

Solve quadratic equations to find the unknown numbers.

1 Ben thinks of a number. He adds 3 to it, then squares the result, then subtracts 16 times the number he first thought of. His answer is 0. What numbers could he have been thinking of?

2 Emily thinks of a number. She squares it, then multiplies the answer by 4, then subtracts 16 times the number she first thought of. Her answer is 84. What numbers could she have been thinking of?

3 Dan thinks of a positive even number. He squares it, and then adds 6 times his positive even number. His answer is 352. What positive even number was he thinking of?

4 Two positive numbers have a difference of 5 and a product of 176. Calculate the two numbers.

5 The product of two positive consecutive odd numbers is added to 4 times the smaller odd number. The answer is 135. Calculate the two numbers.

6 Piri thinks of a positive even number. He adds 5 to it, then squares the result. He then subtracts 16 times his positive even number. The answer is 137. What number was he thinking of?

7 Angus thinks of a positive odd number. He squares it, then adds 4 times the number he first thought of. The answer is 165. What number was he thinking of?

8 Amber is 3 years younger than her older sister, and 2 years older than her younger sister. The product of Amber's age and her younger sister's age is subtracted from the product of Amber's age and her older sister's age. The answer is 40. How old is Amber?

9 Peter thinks of a positive even number. He adds 5 to it, then he squares the result. He finds the answer is the same as multiplying his original number by 4 and adding 137. What number was he thinking of?

10 Liam is the youngest of three brothers who were born at two-year intervals. The product of the ages of the two youngest brothers is the same as 4 times the oldest brother's age, plus 83. Calculate the ages of the three brothers.

11 Jacob's dad is 25 years older than he is. Jacob multiplies their ages and then subtracts 10 times his father's age. The answer is 74. How old are they?

12 Wiremu thinks of a positive number. He adds 5 to it, then squares his answer. He calls the result A. He then subtracts 1 from his original positive number, and squares the answer. He calls the result B. He subtracts B from A and gets 60. What number was he thinking of?

ISBN: 9780170354226

b Problems about lengths, areas and volumes

Hints:

- If you are not given one, do your own sketch and add all the information to it.
- Sometimes you might be given an inequality. Treat it as an equality (= sign) until the end, then check which side of your answer complies with the question.

Example: The length of the shorter of the parallel sides of a trapezium is half its height. The longer of the parallel sides is 6 cm longer than the shorter of the parallel sides. The area of the trapezium is 140 cm^2. What is the height of the trapezium?

$$\begin{aligned}\text{Area} &= \frac{a+b}{2} \times h \\ &= \frac{x+(x+6)}{2} \times 2x \\ &= \frac{2x+6}{2} \times 2x \\ &= (x+3) \times 2x \\ &= 2x^2 + 6x\end{aligned}$$

$$\therefore 2x^2 + 6x = 140$$

$$2x^2 + 6x - 140 = 0$$

$$x^2 + 3x - 70 = 0$$

$$(x+10)(x-7) = 0$$

$$\therefore x = -10 \text{ or } x = 7$$

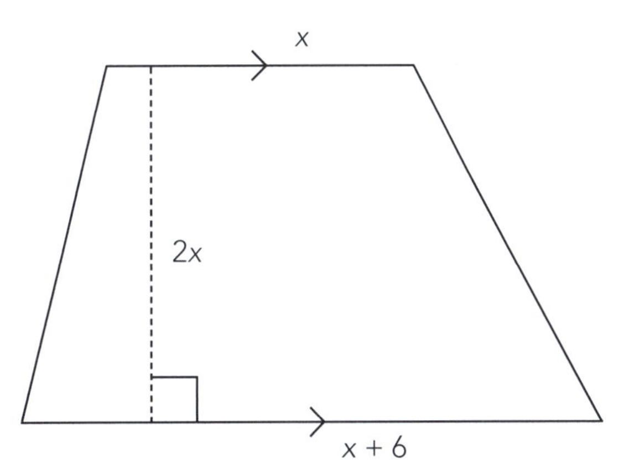

A length has to be positive, so -10 is not a possibility, therefore $x = 7$, and the height is 14 cm.
For the area to be **less** than 140 cm^2, the height must be **less** than 14 cm.

Try for yourself.

The top and bottom of a parallelogram are both 3 cm longer than its vertical height. Its sides are twice as long as its vertical height. If its area is less than 88 cm^2, calculate its vertical height.

ISBN: 9780170354226

Solve quadratic equations to find the unknowns.

1 The top and bottom of a rectangle are 4 cm longer than its sides. The area of a rectangle is less than 165 cm^2. Calculate the maximum length of the sides.

2 The base of a triangle is 7 cm longer than its height. Its area is more than 72 cm^2. Calculate the minimum height of the triangle.

3 The longest side of a right-angled triangle is 8 cm longer than the shortest side. The third side is 7 cm longer than the shortest side. Use Pythagoras' Theorem to calculate the lengths of the three sides.

4 A rectangular piece of fabric has a square cut out of one corner. The longer edge of the piece of fabric is 11 cm longer than the side of the square, and the shorter edge is 3 cm longer than the side of the square. If the remaining area is103 cm^2, calculate the dimensions of the original rectangular piece of fabric.

5 A cuboidal box is 8 cm high. Its length is 3 less than twice its width. Its volume is 280 cm^3. Calculate the dimensions of the box.

ISBN: 9780170354226

Hint: You will need to solve these last four questions by using the quadratic formula or on your graphics calculator.

6 Strips of garden are to be created along two sides of a paved area. The entire area is a rectangle which is 25 m long and 15 m wide. The strips of garden are x m wide. The owners can afford only 105 m^2 of paving. How wide are the garden strips?

7 Another garden is to be created, but it is to have a paved path around the edge. The garden has to be half the total area, and all the strips of path are x m wide. Calculate the width of the path.

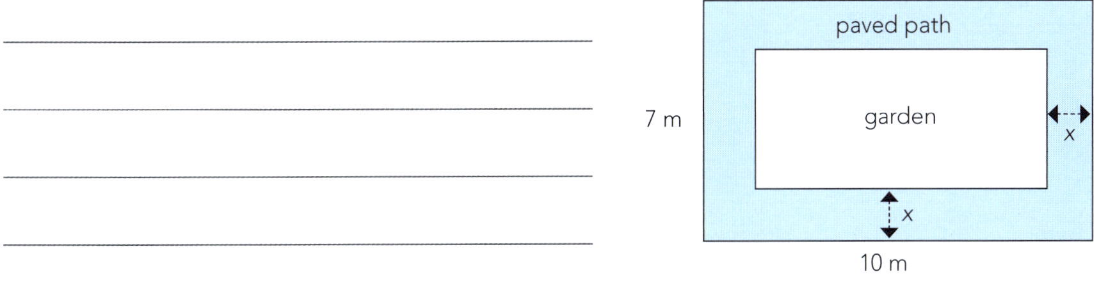

8 This garden, which is 12 m by 7.5 m, has a path on three sides. If the paved area must be a total of 45 m^2, calculate the width of the paths.

9 This piece of garden has just one strip of path beside it. If the **entire** area must be 69 m^2, calculate the width of the path.

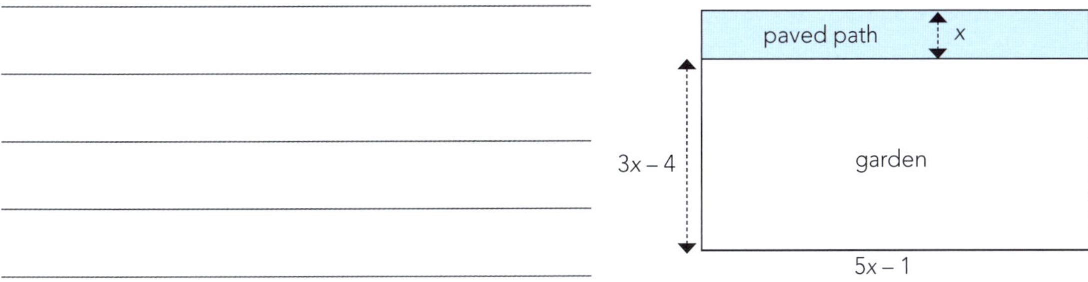

2 Given information about the graph

a $y = ax^2$ and $y = -ax^2$

You need to consider two factors:

1 Sign: $y = +x^2$ $y = -x^2$

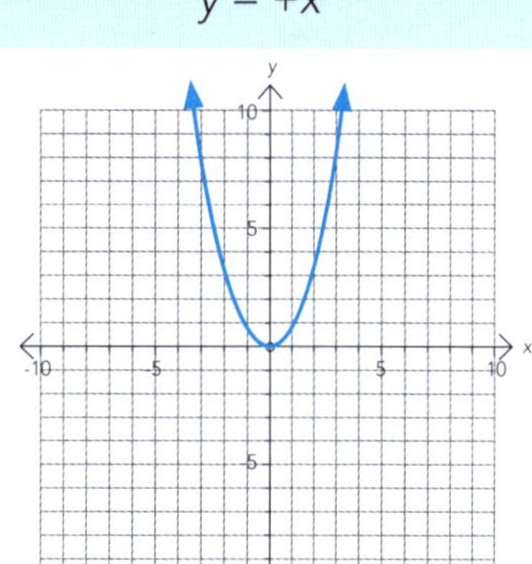

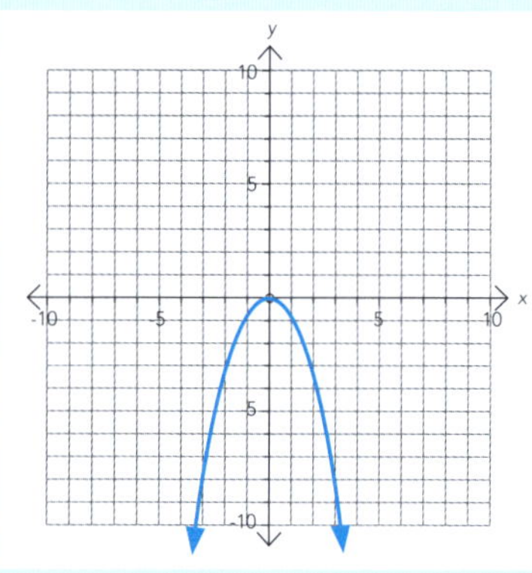

2 The effect of ***a***

***a* > 1**, e.g. $y = 2x^2$ ***a* < 1**, e.g. $y = \frac{1}{2}x^2$

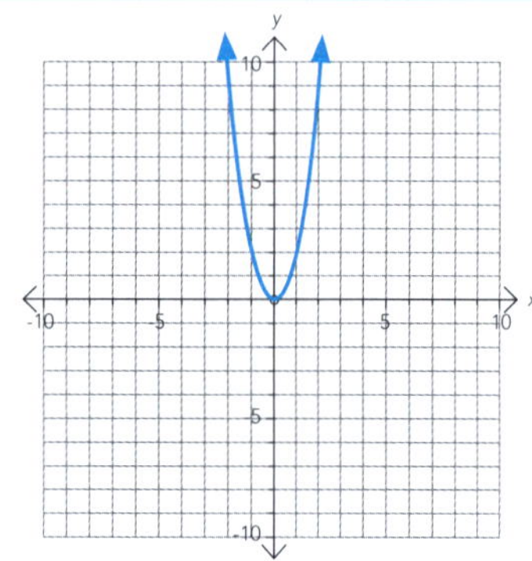

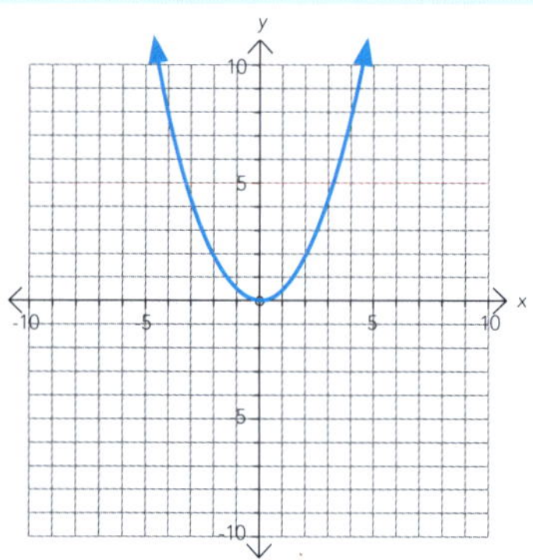

You will be given a point on the graph so that you can calculate the size of ***a***.

Examples:

1 A quadratic function takes the form $y = ax^2$. Calculate its equation if the point (2, 12) lies on the graph.

When $x = 2$, $y = 12$, so $12 = a(2)^2$

$$a = \frac{12}{2^2} = 3$$

$\therefore$ equation is $y = 3x^2$

2 A quadratic function takes the form $y = ax^2$. Calculate its equation if the point (2, -1) lies on the graph.

When $x = 2$, $y = -1$, so $-1 = a(2)^2$

$$a = -\frac{1}{2^2} = -\frac{1}{4}$$

$\therefore$ equation is $y = -\frac{1}{4}x^2$

  ISBN: 9780170354226

Try these questions.

1 A quadratic function takes the form $y = ax^2$. Calculate its equation if the point (2, 2) lies on the graph.

2 A quadratic function takes the form $y = ax^2$. Calculate its equation if the point (1, -2) lies on the graph.

3 A quadratic function takes the form $y = ax^2$. Calculate its equation if the point (2, -1) lies on the graph.

4 A quadratic function takes the form $y = ax^2$. Calculate its equation if the point (5, 5) lies on the graph.

5 A quadratic function takes the form $y = ax^2$. Calculate its equation if the point (2, -6) lies on the graph.

6 A quadratic function takes the form $y = ax^2$. Calculate its equation if the point (-3, -45) lies on the graph.

ISBN: 9780170354226

b $y = ax^2 + c$

In addition to what you learned above, you need to consider the effect of c, which will shift the graph up or down.

c > 1, e.g. $y = -2x^2 + 3$ **c < 1**, e.g. $y = \frac{1}{2}x^2 - 4$

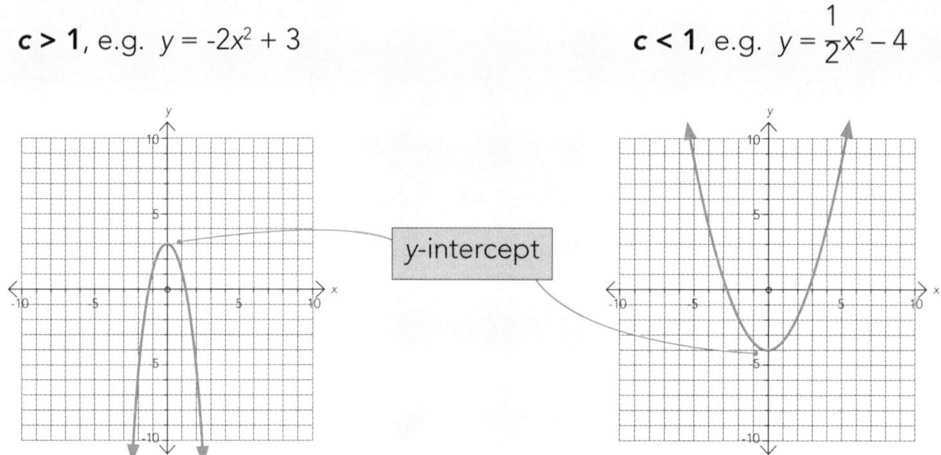

You will be told the *y*-intercept and a point on the graph so that you can calculate the size of ***a***.

Examples:

1 A quadratic function takes the form $y = ax^2 + c$. Calculate its equation if the point (2, 7) lies on the graph, and the *y*-intercept is 5.

When $x = 2$, $y = 7$, so

$$7 = a(2)^2 + 5$$
$$4a = 2$$
$$a = \frac{1}{2}$$
$$\therefore \text{ equation is } y = \frac{1}{2}x^2 + 5$$

Sometimes you will need to sketch the graph in order to find the coordinates of points.

2 An archway is modelled by a quadratic function where *y* is the height above the ground and *x* is horizontal distance from the centre of the archway. The archway is 8 m high. At 6 m above the ground it is 4 m wide. Calculate the equation of the model.

The sketch tells you that the equation passes through the points (0, 8) and (2, 6).

When $x = 2$, $y = 6$, so

$$6 = -a(2)^2 + 8$$
$$-4a = -2$$
$$a = \frac{1}{2}$$
$$\therefore \text{ equation is } y = \frac{1}{2}x^2 + 8$$

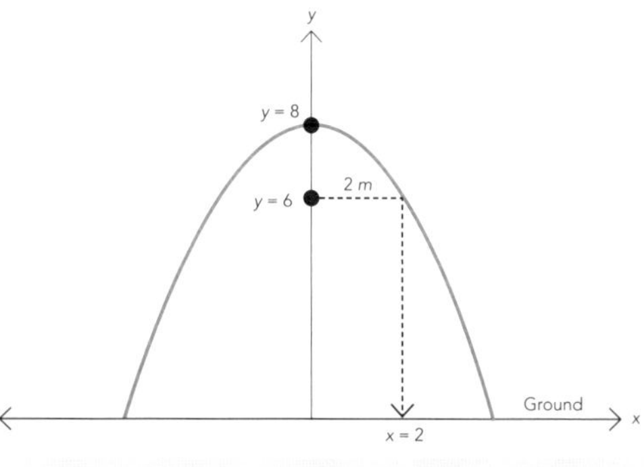

ISBN: 9780170354226

Try these questions.

1 A quadratic function takes the form $y = ax^2 + c$. Calculate its equation if the point (2, 13) lies on the graph, and the y-intercept is 7.

2 A quadratic function takes the form $y = ax^2 + c$. Calculate its equation if the graph passes through the points (-2, -10) and (0, 2).

3 A quadratic function takes the form $y = ax^2 + c$. Calculate its equation if the graph passes through the points (-10, -20) and (0, -10).

4 An archway is modelled by a quadratic function where y is the height above the ground and x is horizontal distance from the centre of the archway. The archway is 6 m high. At 1 m above the ground, it is 10 m wide. Calculate the equation of the model.

5 A canal is modelled by a quadratic function where y is its depth below the ground and x is the horizontal distance from the centre of the canal. The canal is 7 m deep. Where it is 8m wide, it is 3 m deep. Calculate the equation of the model.

ISBN: 9780170354226

c $y = a(x + d)(x - d)$

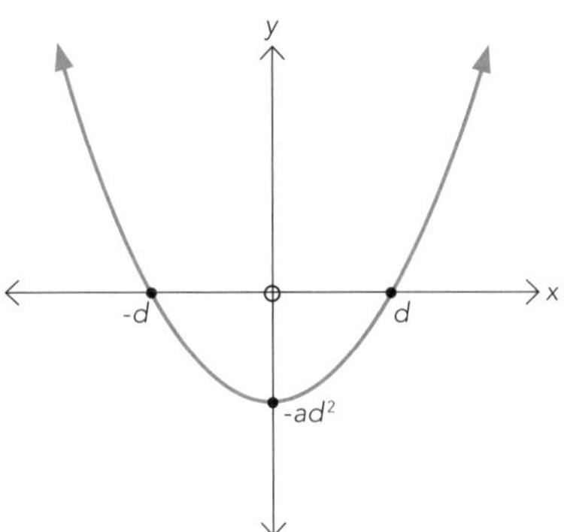

For equations in this form:

- The x-intercepts are $-d$ and $+d$.
- The y-intercept is $\boldsymbol{a \times d \times -d = -ad^2}$
- The parabola will be symmetrical about the y-axis.
- If you are not given a picture, it is a good idea to draw one.

Example: The width of a canal at ground level is 12 m. The sides of the canal can be modelled by a quadratic expression that would give a maximum depth of 14.4 m. However, the base of the canal is flat and this has a width of 8 m. Let x be the distance from the centre of the arch and y its height above the ground.

a Find the equation used to model the sides of the canal.

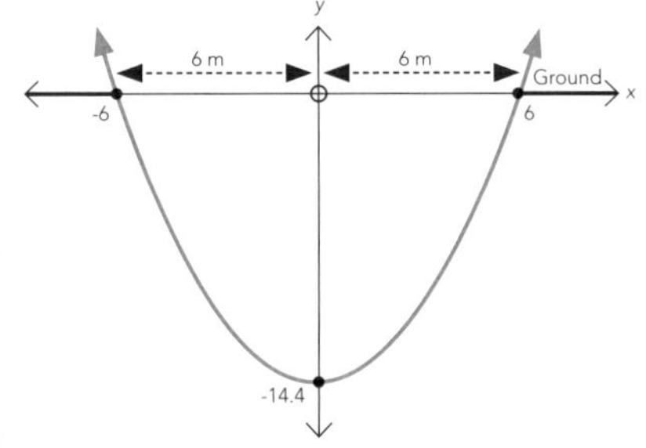

$y = a(x + d)(x - d)$

Width = 12 m $\longrightarrow d = 6$

$\therefore y = a(x + 6)(x - 6)$

Maximum depth = 14.4

$\therefore -ad^2 = -a \times 6 \times 6 = -14.4$

$$a = \frac{-14.4}{-36} = 0.4$$

So the equation is $y = 0.4(x + 6)(x - 6)$.

b Calculate the depth of the canal.

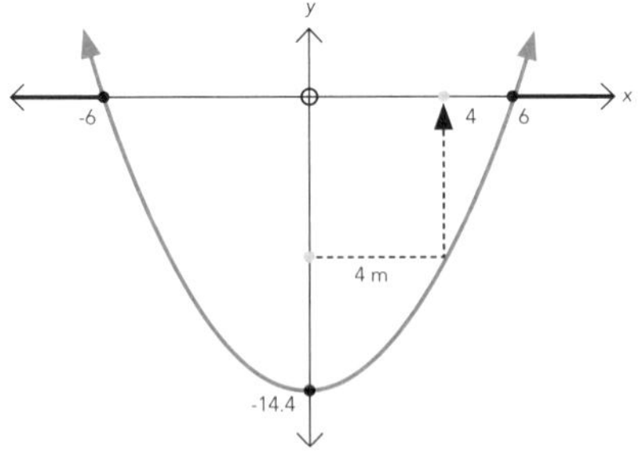

Width of the base = 8 m $\longrightarrow x = 4$

$\therefore y = 0.4(4 + 6)(4 - 6)$

$y = 0.4 \times 10 \times -2$

$y = -8$

So the canal is 8 m deep.

c The bottom of the canal has to be dug out in order to make it 14 m deep. Calculate the width of its new base.

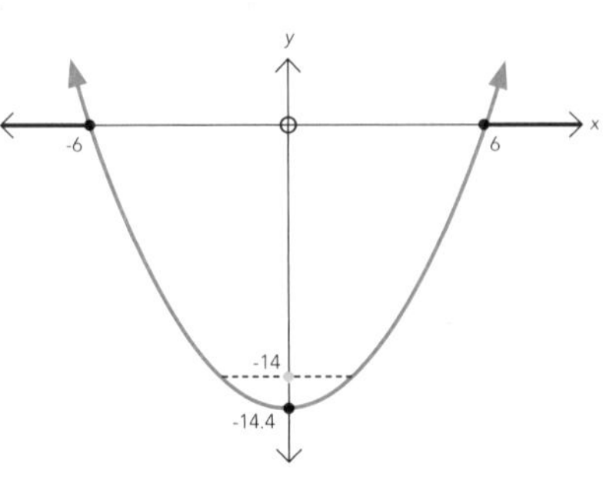

$$-14 = 0.4(x + 6)(x - 6)$$

$$(x + 6)(x - 6) = -\frac{14}{0.4} = -35$$

$$x^2 - 36 = -\frac{14}{0.4} = -35$$

$$x^2 = 1 \longrightarrow x = \pm 1$$

$\therefore$ The new base must be 2 m wide.

ISBN: 9780170354226

Try these questions.

1 The width of a canal at ground level is 8 m. The sides of the canal can be modelled by a quadratic expression that would give a maximum depth of 12 m. However, the base of the canal is flat and this has a width of 4 m. Let x be the distance from the centre of the canal and y its depth below the ground. Calculate the equation of the model and the depth of the canal.

2 An arch is 8 m wide and 8 m high. Let x be the distance from the centre of the arch and y its height above the ground. The shape of the arch is modelled by a quadratic expression. The arch has a horizontal bar which is 4 m long. Calculate the height of the bar.

3 An arched roof is 20 m wide and 4 m high. Let x be the distance from the centre of the arch and y its height above the top of the walls. The shape of the arched roof is modelled by a quadratic expression. The roof has a horizontal brace that is 3 m above the top of the walls. Calculate the width of the brace.

4 A feature in a skateboard park is parabolic in shape, so it can be modelled by a quadratic equation. Let x represent the distance from its centre and y its depth below the surrounding ground. The feature is 30 m wide and at its lowest point is 2.25 m deep. During a severe rainstorm, a puddle forms in the bottom. The surface of the puddle is 2 m below the surrounding ground level. Calculate the width of the puddle.

ISBN: 9780170354226

d $y = a(x - d)(x - e)$

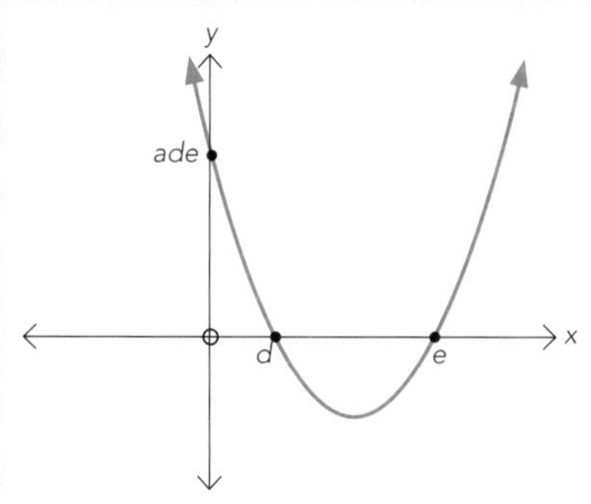

For equations in this form:

- The x-intercepts are d and e.
- The y-intercept is $\boldsymbol{a \times -d \times -e = ade}$.
- The parabola will **not** be symmetrical about the y-axis.

Example: A 6 m wide arch has a maximum height of 4.5 m. It is located so that its centre is 5 m from a fence. The archway can be modelled by a quadratic function where y is the height above the ground and x is the distance from the fence.

a Calculate the equation that models the archway.

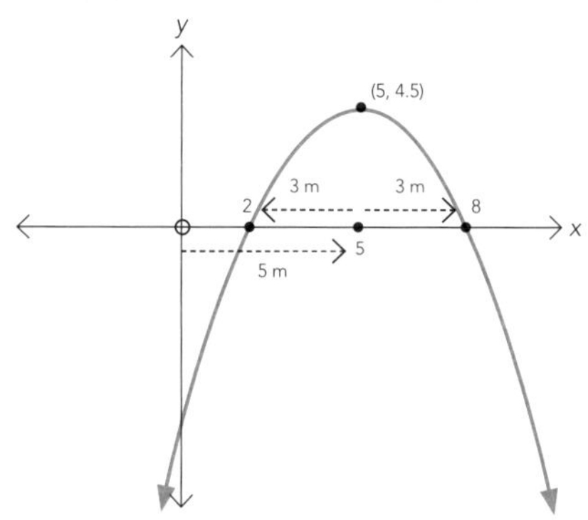

$y = a(x - d)(x - e)$

From the diagram, the x-intercepts are at $x = 2$ and $x = 8$

$\therefore y = a(x - 2)(x - 8)$

Its centre is at $x = 5$, maximum height = 4.5 ⟶ passes through (5, 4.5)

$\therefore 4.5 = a(5 - 2)(5 - 8)$

$4.5 = a \times 3 \times -3$

$a = \dfrac{4.5}{-9} = -0.5$

So the equation is $y = -0.5(x - 2)(x - 8)$.

b A horizontal string of lights is to be suspended across the archway at two and a half metres above the ground. Calculate the length of the string.

Height of string = 2.5 m ⟶ $y = 2.5$

$\therefore 2.5 = -0.5(x - 2)(x - 8)$

$(x - 2)(x - 8) = \dfrac{2.5}{-0.5} = -5$

$x^2 - 10x + 16 = -5$

$x^2 - 10x + 21 = 0$

$(x - 3)(x - 7) = 0$

$\therefore x = 3$ or $x = 7$

So the lights are strung between $x = 3$ m and $x = 7$ m, therefore the string must be 4 m long.

ISBN: 9780170354226

Try these questions.

1 A 6 m wide arch has a maximum height of 2.25 m. It is located so that its centre is 4 m from a fence. The archway can be modelled by a quadratic function. A 2 m long horizontal string of lights is to be suspended across the archway. Calculate the height of the string above the ground.

2 An 8 m wide arch has a maximum height of 3.2 m. It is located so that its centre is 6 m from a fence. The archway can be modelled by a quadratic function. A horizontal string of lights is to be suspended across the archway at 3 m above the ground. Calculate the length of the string.

3 A pond, whose sides can be modelled by a quadratic function, is constructed. It is 16 m wide and at its centre it has a depth of 3.2 m. Let y represent its depth in metres and x the distance in metres from its edge. How wide is the surface of the water in the bottom when it is filled to a depth of 20 cm?

4 A rugby goal-kicker attempts a goal from directly in front of the goal posts. The flight of the ball can be modelled by a quadratic function, and it reaches its highest point at the goal posts. Let x be the distance from the point at which he begins his run. He kicks the ball 24 m from the goal posts.

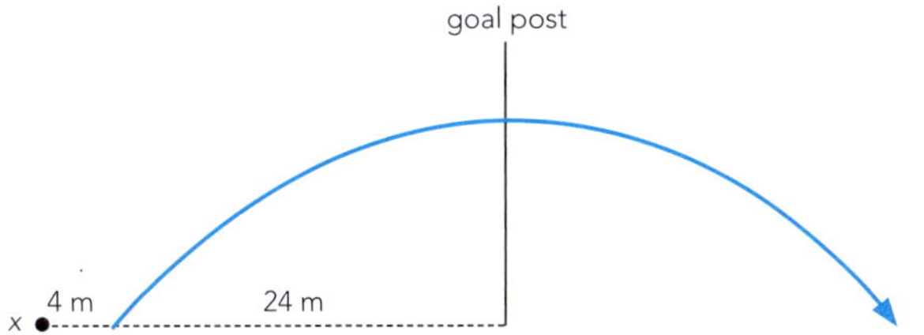

When the ball is 20 m from where he kicked it, it is 2.8 m high. The goal is 3 m high. Will the ball go over the goal? Justify your answer with calculations.

ISBN: 9780170354226

Rearranging expressions

1 Where the subject appears once

- Get rid of fractions by multiplying by the lowest common multiple of the denominators.
- Use normal equation-solving rules to isolate the subject.

Examples:

1 Make m the subject of $y = mx + c$

$$y - c = mx$$
$$\frac{y-c}{x} = m$$
$$m = \frac{y-c}{x}$$

2 Make a the subject of $A = \frac{a+b}{2}h$

$$2A = h(a+b)$$
$$\frac{2A}{h} = a + b$$
$$a = \frac{2A}{h} - b$$

Try these yourself.

1 Make r the subject of $A = \pi r^2$

2 Make b the subject of $b^2 - 4ac = x$

3 Make c the subject of $b^2 - 4ac = x$

4 Make r the subject of $S = \frac{4}{3}\pi r^3$

5 Make L the subject of $n\lambda = \frac{dx}{L}$

6 Make y the subject of $E = \frac{1}{2}ky^2$

7 Make V the subject of $F = \frac{mV^2}{r}$

8 Make v the subject of $\frac{1}{f} = \frac{1}{u} + \frac{1}{v}$

ISBN: 9780170354226

2 Where the subject appears more than once

Example: Make x the subject of $y = \frac{x}{x+2}$

Step 1: Get rid of fractions by multiplying by the denominator: $xy + 2y = x$

Step 2: Collect x terms on the left and others on the right: $xy - x = -2y$

Step 3: Factorise the left side with x outside the brackets: $x(y - 1) = -2y$

Step 4: Divide so x becomes the subject: $x = \frac{-2y}{y-1}$ or $\frac{2y}{1-y}$

Try these yourself.

1 Make x the subject of $y = \frac{x+1}{x}$

2 Make x the subject of $y = \frac{3x}{x+1}$

3 Make x the subject of $y = \frac{3x-2}{x}$

4 Make x the subject of $y = \frac{2x+1}{3x-2}$

5 Make x the subject of $y = \frac{x-1}{x+2}$

6 Make x the subject of $\frac{y}{5} = \frac{x}{x-3}$

3 Where there is a root sign

Example: Make x the subject of $y = 4 - \sqrt{2x + 1}$

Step 1: Reorganise so that only the root term is on the left: $\sqrt{2x+1} = 4 - y$

Step 2: Square both sides: $2x + 1 = 16 - 8y + y^2$

Step 3: Reorganise to make x the subject: $2x = 15 - 8y + y^2$

$$x = \frac{15 - 8y + y^2}{2}$$

Try these yourself.

1 Make x the subject of $y = \sqrt{x - 3}$

2 Make x the subject of $y = 3 + \sqrt{4x - 1}$

3 Make x the subject of $y = \sqrt{x - 3} + 2$

4 Make x the subject of $y = \sqrt{\frac{x}{x + 1}}$

5 Make x the subject of $y = \sqrt{\frac{x - 2}{x + 4}}$

6 Make k the subject of $T = 2\pi \sqrt{\frac{m}{k}}$

7 Make c the subject of $x = \frac{-b + \sqrt{b^2 - 4ac}}{2a}$

8 Make L the subject of $f = \frac{1}{2\pi\sqrt{LC}}$

ISBN: 9780170354226

Practice questions

Practice question one

a Simplify these.

i $\dfrac{(2x^2)^4}{3\,(4x^5)^2}$

ii $(4x^{\frac{1}{2}})^{\frac{3}{2}}$

iii $\sqrt{\dfrac{(4x^{\frac{1}{2}})^{\frac{3}{2}}}{x^{-\frac{1}{4}}}}$

b Tama's teacher asks him how many Merit grades he earned in Mathematics in Year 11. Tama set his teacher this puzzle. If you

- add 5 to the number of Merit grades
- square the result
- subtract 9 times the number of Merit grades
- subtract 37

then the answer will be 0.

How many Merit grades did Tama earn in Mathematics?

c A decorative archway is suspended at its highest point from a crossbeam that is 3 m long. The crossbeam is supported by a post that is 8 m high. The archway can be modelled by a quadratic function where y is the height above the ground and x is the distance from the post.

i Give the equation of the model if the archway is 4 m wide at ground level.

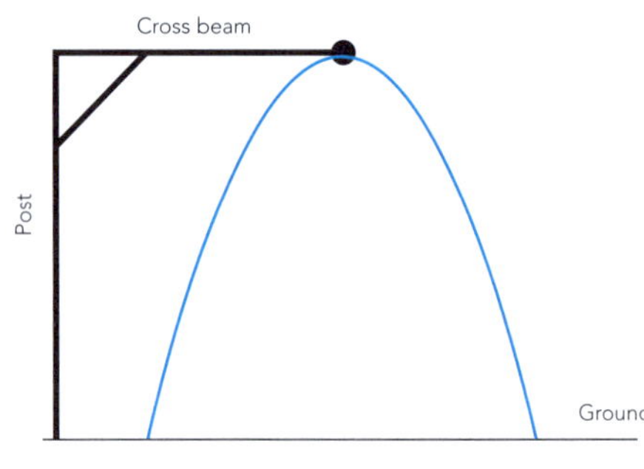

ii At what height does the archway have a width of 2 m?

Practice question two

a **i** Factorise $10x^2 + 7x - 12$

ii Solve $10x^2 + 7x - 12 = 0$

b **i** Simplify $\dfrac{2x^2 - 18}{(x - 3)(x + 1)}$

 ISBN: 9780170354226

ii Solve $\dfrac{2x^2 - 18}{(x - 3)(x + 1)} = x$

c Find the possible values for k if no real roots exist for $2x^2 + (2k - 1)x + (k + 3) = 0$.

d The length of the base of an isosceles triangle is $(4x - 6)$ cm.

If its area is $(2x^2 - x - 3)$ cm^2, find an expression for the length of side L.

ISBN: 9780170354226

Practice question three

a Solve these equations.

i $\log_x 32 = 5$

ii $\log x = 5\log 2$

b Evaluate $\log_2 32 - \frac{1}{2}\log_2 16$

c Solve $\frac{4(27^x)}{3^{4x-1}} = 36$

d Two possums, a male and a female, are released into a large area that is otherwise possum free. It is known that the rate of growth of a possum population is 40% per year. If no factors prevent the population growing at this rate:

i Write expressions for the size of the possum population after *n* years.

 ISBN: 9780170354226

ii How many more possums will there be after 10 years?

iii After how many years will there be 20 possums in the area?

iv The population scientist uses this formula when doing some calculations involving the possum population: $N = \frac{\log 10}{\log(\frac{100 + m}{100})}$

Explain what she could be calculating, and state what the variables N and m stand for.

ISBN: 9780170354226

Answers

All answers are rounded to a maximum of 4 dp. Professional judgement should apply.

Powers (pp. 7–18)

Multiplying powers (p. 7)

1 b^5 2 y^{14} 3 $30b^{15}$
4 y^8 5 $8z^{10}$ 6 $y^{1.5}$
7 $-6y^6$ 8 $y^{5.6}$ 9 $9y^7$
10 $-15y^{2.4}$

Dividing powers (p. 8)

1 b^{10} 2 y^2
3 $\frac{1}{y^3}$ or y^{-3} 4 $5b^5$
5 $\frac{1}{2y^5}$ or $0.5y^{-5}$ 6 $\frac{x^2y}{z}$ or x^2yz^{-1}
7 $\frac{1}{3y^2z}$ or $0.\dot{3}y^{-2}z^{-1}$ 8 $\frac{3x}{y}$ or $3xy^{-1}$
9 $-3x$ 10 $\frac{2x^2z^2}{y^3}$ or $2x^2y^{-3}z^2$

Powers of powers (p. 9)

1 b^{24} 2 $16x^{24}$
3 a^{14} 4 $-y^{15}$
5 $9x^{10}$ 6 $-64x^{12}$
7 $-50x^6y^8$ 8 $216x^6y^{12}z^9$
9 $-64a^6b^4c^8$ 10 $18x^2y^6z^4$
11 $\frac{x^6}{125}$ or $\frac{1}{125}x^6$ 12 $\frac{9x^8}{64}$ or $\frac{9}{64}x^8$

Putting it together so far (p. 10)

1 b^{10} 2 1
3 b^2c^3 4 $\frac{25}{4b}$ or $6.25b^{-1}$
5 $\frac{a^6}{8}$ or $0.125a^6$ 6 $\frac{16z}{49}$ or $0.3265z$
7 $\frac{27}{z^{1.5}}$ or $27z^{-1.5}$ 8 $\frac{4z^6}{25x^2y^4}$ or $0.16x^{-2}y^{-4}z^6$
9 $\frac{4a^2}{b^4}$ 10 $\frac{-a^6}{64b^3}$

Negative powers (pp. 11–12)

1 With numbers

1 $\frac{1}{9}$ 2 $\frac{5}{4}$ 3 4
4 $\frac{27}{8}$ 5 $\frac{5}{9}$ 6 $\frac{9}{16}$
7 $\frac{16}{25}$ 8 $\frac{8}{125}$ 9 16
10 $\frac{1}{81}$ 11 $-\frac{1}{64}$ 12 $\frac{25}{49}$

2 With variables

1 $\frac{1}{b^{12}}$ 2 $\frac{5}{a^2}$ 3 $\frac{a}{b^2}$
4 $\frac{8a^4}{b^3}$ 5 $\frac{1}{a^2b^3}$ 6 $\frac{1}{9a^2b^{10}}$
7 $\frac{2}{a^6b^{12}c^3}$ 8 a^3 9 $a^{10}b^2$
10 $\frac{3a^2b^4}{c^2}$ 11 $\frac{b^2}{25a^4}$ 12 $\frac{a^2}{4bc^3}$

Fractional powers — surds (pp. 13–14)

Write in index form

1 $a^{\frac{1}{6}}$ 2 $a^{\frac{3}{4}}$ 3 $a^{\frac{1}{2}}$
4 a^8 5 $9a^2$ 6 $3a^{\frac{5}{4}}$

Write in surd form

1 $\sqrt[3]{a}$ 2 $\sqrt[5]{a^2}$ 3 $8\sqrt[3]{a}$
4 $2\sqrt[3]{a}$ 5 $12\sqrt[7]{a^5}$ 6 $9\sqrt{a^3}$

Calculations involving surds

1 a^5 2 a^3
3 $a^{\frac{9}{4}}$ or $\sqrt[4]{a^9}$ 4 $a^{\frac{8}{5}}$ or $\sqrt[5]{a^8}$
5 a^6 6 $a^{\frac{1}{2}}$ or $\sqrt{a}$
7 $a^{\frac{3}{8}}$ or $\sqrt[8]{a^3}$ 8 $a^{\frac{6}{5}}$ or $\sqrt[5]{a^6}$

Combined roots, powers and coefficients (p. 15)

1 $2a^2$ 2 $4a^3$
3 $9a^{\frac{3}{5}}$ or $9\sqrt[5]{a^3}$ 4 $7a^{\frac{11}{2}}$ or $7\sqrt{a^{11}}$
5 $4a^{\frac{1}{3}}$ or $4\sqrt[3]{a}$ 6 $10a^4$
7 $\frac{2}{5}x^{\frac{1}{6}}$ or $\frac{2}{5}\sqrt[6]{x}$ 8 $24x^3$

Powers where the base is a number (p. 16)

1 2^{3n-3} or 8^{n-1} 2 3^{3-2n}
3 2^n 4 $\frac{1}{2^n}$
5 3×2^{9n} 6 2^{n-1}
7 $\frac{1}{3}$ 8 2^{2n+4} or 4^{n+2}

Powers — putting it all together (pp. 17–18)

1 $9x^{14}$ 2 $4x^{11}$

ISBN: 9780170354226

3 $6x^2$ 4 $\frac{1}{3x^5}$
5 $\frac{7}{5x^2}$ 6 $\frac{3}{x^2}$
7 $16x^{10}$ 8 $\frac{3}{8x^9}$
9 $\frac{1}{x^2}$ 10 x
11 $2x^5y$ 12 $\frac{1}{10a^4}$
13 $\frac{1}{4x^4}$ 14 $\frac{xy^3}{2}$ or $\frac{1}{2}y^3x$
15 $\frac{5}{8x^4}$ 16 $\frac{1}{2x^{10}}$
17 $\frac{x^3y^2}{2}$ 18 $\frac{x^4}{3y^3}$

Logarithms (pp. 19–35)

Logarithm introduction (pp. 19–20)

Write in log form

1 $\log_2 32 = 5$ 2 $\log_{10} 10\,000 = 4$
3 $\log_a c = b$ 4 $\log_{10} 10 = 1$
5 $\log_5 125 = 3$ 6 $\log_3 81 = 4$

Write in exponent form

1 $4^2 = 16$ 2 $10^2 = 100$ 3 $2^6 = 64$
4 $5^4 = 625$ 5 $6^2 = 36$ 6 $3^3 = 27$

Write in log form

1 $\log_{13} 1 = 0$ 2 $\log_5 5 = 1$
3 $\log_{10} 0.001 = -3$ 4 $\log_9 3 = \frac{1}{2}$
5 $\log_4 \frac{1}{4} = -1$ 6 $\log_{10} 0.01 = -2$
7 $\log_{16} \frac{1}{4} = -\frac{1}{2}$ 8 $\log_{25} 5 = 0.5$ or $\frac{1}{2}$
9 $\log_{125} 5 = \frac{1}{3}$ 10 $\log_8 \frac{1}{2} = -\frac{1}{3}$

Write in exponent form

1 $15^0 = 1$ 2 $10^{-3} = \frac{1}{1000}$
3 $36^{\frac{1}{2}} = 6$ 4 $16^1 = 16$
5 $32^{\frac{1}{5}} = 2$ 6 $10^{-1} = 0.1$
7 $8^0 = 1$ 8 $2^{-1} = \frac{1}{2}$
9 $49^{-\frac{1}{2}} = \frac{1}{7}$ 10 $1000^{-\frac{1}{3}} = \frac{1}{10}$

11

Log	Value
$\log_2 200$	7.645
$\log_{10} 4000$	3.602
$\log_5 1.5$	0.25
$\log_7 300$	2.93
$\log_{100} 240$	1.19
$\log_6 10\,000$	5.1404
$\log_3 100$	4.192
$\log_3 960$	6.251

Logarithm rules (pp. 21–23)

1 $\log a + \log b = \log ab$

1 log 12 2 log 60
3 log 90 4 log 8
5 log 1 6 log pqr
7 log 6 + log 1 or log 2 + log 3 8 log 9 + log 1 or log 3 + log 3
9 log p + log q 10 log a + log b + log c

2 $\log a - \log b = \log \frac{a}{b}$

1 log 6 2 log 5
3 log 6 4 log 4
5 $\log \frac{b}{c}$ 6 $\log \frac{fg}{h}$
7 log 15 – log 3 8 log 20 – log 1
9 log 7 + log 4 – log 2 10 log n – log p – log q

3 $n\log a = \log a^n$

1 log 25 2 log 8
3 log 1 000 000 4 log 7
5 log 5 6 $\log \sqrt[a]{b}$

Putting logarithm rules together (pp. 22 - 23)

1 4 2 8log 2 3 log 7
4 2 5 2log 10 6 $\frac{3}{2}$
7 2log 5 8 log 2 9 $\frac{5}{2}$
10 0 11 $\frac{1}{4}$ 12 $\frac{1}{2}$
13 log 10 14 0

Solving logarithm (exponential) equations (pp. 24–34)

1 Basic

1 16 2 128 3 5
4 1 5 5 6 $\frac{1}{8}$
7 $\frac{1}{9}$ 8 $\frac{1}{5}$

2 Finding the base

1 10 2 3 3 23
4 16 5 $\frac{1}{6}$ 6 Any value
7 125 8 100

3 Finding the exponent

1 4 2 4 3 0
4 3 5 1 6 $\frac{1}{2}$
7 $\frac{1}{2}$ 8 $-\frac{1}{6}$

4 Finding the exponent when you don't have 'nice' numbers

1 2.1962 2 6.6439
3 0.1505 4 -0.9125
5 1.6826 6 2.6990

ISBN: 9780170354226

7 0.6425
8 -1.7304
9 0.3155
10 0.1505
11 0.8135
12 1
13 2.7944
14 -1.5

Applications of exponential equations

1 a $A = 500(1.06)^n$
b Amount owed = $595.51
c Less than 5.7745 years

2 a $A = 3500(0.85)^n$
b The car will be worth $1552.97
c Maximum time is 3.443 years

3 a Kiri: $A = 5000(1.05)^n$, Jack: $A = 5000(1.03)^{2n}$
b Kiri: $6381.41, Jack: $6719.58
c Kiri: 14.21 years, Jack: 11.72 years.
d Time = $\frac{\log 2}{\log(\frac{100+x}{100})}$ years

4 a $A = 2(1.24)^n$
b Area = 3.81 m^2
c After 7.482 years

5 a $A = 500\,000(1.15)^n$
b 374 503 bacteria
c The increase in the number of bacteria n days since the start
d After 6.264 days

6 a $A = 50\,000(0.82)^n$
b 34 800 bees
c After 11.60 weeks

7 a Worth $1370.24
b Worth an extra $507.32
c The difference between investing the money for p years and $(p + q)$ years
d After 22 years

8 a $A = 35\,000(0.98)^n$
b 7535 L
c After 34.31 days

9 a Value = $1050(0.65)^n$
b Worth $288.36
c Worth $166.53
d The difference between the value of the phone after p years and $(p + q)$ years
e After 7.0674 years

Fractions (pp. 36–42)

Multiplying and dividing fractions (pp. 36–37)

1 $\frac{8}{5}$ or 1.6
2 $\frac{4y^2}{5}$ or $\frac{4}{5}y^2$
3 $\frac{2}{25b^2}$
4 $6x$
5 $8x$
6 $\frac{z^5}{y^2}$
7 $\frac{3x}{4y}$
8 $3x$
9 $\frac{10}{3a}$
10 $\frac{xy}{4}$
11 $\frac{16}{3x^2}$
12 $\frac{36y}{x}$
13 $\frac{24x^3}{y^2}$
14 $\frac{1}{2x^2}$
15 $\frac{1}{20y}$
16 $\frac{y^2}{8x}$
17 $\frac{x^2}{3}$
18 $\frac{125x}{64y^9}$

Adding and subtracting fractions (pp. 38–39)

1 $\frac{7}{y}$
2 $\frac{2y+6x}{xy}$
3 $\frac{3y-2x}{12}$
4 $\frac{10-x^2y}{5xy}$
5 $\frac{5xy-3}{15}$
6 $\frac{24-x}{6}$
7 $\frac{1-2xy}{xy}$
8 $\frac{3y+xy}{x^2y}$
9 $\frac{3+xy^2}{3y^2}$
10 $\frac{4xy^2+3}{x^2y}$
11 $\frac{yz^2+x^2z+xy^2}{xyz}$
12 $\frac{3y+3x-2y^2}{3xy}$
13 $\frac{4x-1}{x(x-1)}$
14 $\frac{7x-14}{(x-4)(x+3)}$
15 $\frac{-x^2+4x+2}{(x-3)(x+2)}$
16 $\frac{2x^2-9x+2}{(x+4)(x-3)}$

Solving equations involving fractions (p. 40)

1 $\frac{10}{9}$
2 $\frac{4}{7}$
3 $\frac{14}{5}$
4 -10
5 6
6 5

Solving linear inequations (pp. 41–42)

1 $x \geq 7$
2 $x < 13$
3 $x \geq 11$
4 $x \leq \frac{-7}{9}$
5 $x < \frac{4}{17}$
6 $x \leq 2$
7 $x < \frac{60}{13}$
8 $x \leq 9$
9 $x < 2$
10 $x > 1.9$
11 $x > 2.5$
12 $x \leq \frac{1}{6}$
13 $x < 32$
14 $x < -2$

Polynomials (pp. 43–83)

Expanding (pp. 43–46)

1 Two brackets

1 $x^2 + x - 12$
2 $x^2 - 9x + 14$
3 $x^2 + 13x + 40$
4 $x^2 + 16x + 64$
5 $2x^2 - 8$
6 $6x^2 + 12x - 18$
7 $9x^2 - 30x + 25$
8 $9 - 6x + x^2$
9 $8 - 20x - 12x^2$
10 $x^3 - 6x^2 + 3x - 18$
11 $20x^2 + 7x - 6$
12 $28 + 13x - 6x^2$
13 $x^2 - 1$
14 $25 - 4x^2$
15 $16x^2 - 9$
16 $49x^2 + 28x + 4$

2 Three brackets

1 $x^3 + 6x^2 + 11x + 6$
2 $x^3 - 28x - 48$
3 $-x^3 + 3x^2 + 4x - 12$
4 $x^3 + 9x^2 + 27x + 27$
5 $x^3 - 15x^2 + 75x - 125$
6 $8x^3 - 48x^2 + 96x - 64$

ISBN: 9780170354226

7 $x^3 + 6x^2 - 32$ **8** $12x^3 + 4x^2 - 17x + 6$
9 $2x^3 + 23x^2 + 80x + 75$ **10** $4x^3 - 16x^2 + 21x - 9$
11 $x^3 - 4x^2 - 12x$ **12** $2x^3 - 14x^2 + 24x$
13 $x^3 - 4x^2 - 7x + 10$ **14** $6x^3 - 5x^2 - 10x + 6$

Factorising quadratics (pp. 47–51)

1 Where the coefficient of x^2 is 1

1 $(x + 4)(x + 2)$ **2** $(x + 4)(x + 9)$
3 $(x + 7)(x + 1)$ **4** $(x + 6)(x - 2)$
5 $(x + 4)(x + 13)$ **6** $(x - 3)(x - 6)$
7 $(x - 6)(x + 4)$ **8** $(3 + x)(8 + x)$
9 $(x + 8)(x - 8)$ **10** $(1 + x)(1 - x)$
11 $(x - 5)^2$ **12** $(x + 5)(x - 5)$
13 $(3 - 4y)(3 + 4y)$
14 $(7 + x)(7 - x)$ or $-(x - 7)(x + 7)$
15 $-(x + 6)(x - 1)$ **16** $-(x - 2)(x - 3)$
17 $-(x - 4)(x - 2)$ **18** $(5 - 3x)(5 + 3x)$

2 Where the coefficient of x^2 is not 1, but there is a common factor

1 $4(x + 6)(x - 2)$ **2** $-8(x + 1)(x - 1)$
3 $4(x + 4)(x + 1)$ **4** $8(5 + x)(5 - x)$
5 $-5(x + 1)^2$ **6** $-4(x + 1)(x + 3)$
7 $2(x - 3)(x + 2)$ **8** $5(x + 6)(x - 6)$

3 Where the coefficient of x^2 is not 1, and there is no common factor

1 $(2x + 3)(x + 4)$ **2** $(2x + 1)(x + 6)$
3 $(2x + 3)(2x + 5)$ **4** $(2x + 3)(x - 4)$
5 $(3x - 2)(x - 6)$ **6** $2(2x - 3)(x - 1)$
7 $(x - 5)(4x + 5)$ **8** $2(x - 1)(2x - 1)$
9 $(2x + 3)(5x - 1)$ **10** $(3x - 2)(3x - 1)$
11 $(x + 2)(2 - 3x)$ **12** $2(x + 2)(1 - x)$
13 $(2x + 3)(1 - x)$ **14** $(x - 9)(3x + 2)$
15 $(3x + 2)(4x - 1)$ **16** $(2x + 7)(x - 8)$
17 $(5x + 1)(x - 2)$ **18** $(3x + 1)(2x - 1)$
19 $(3x + 2)(2x + 3)$ **20** $2(x - 3)^2$

Simplifying rational quadratic expressions (p. 52)

1 2 **2** $\frac{1}{4}$ **3** -1
4 $\frac{2}{3}$ **5** $x + 8$ **6** $x - 4$
7 $\frac{x+1}{x-1}$ **8** $\frac{x+2}{x+4}$ **9** $\frac{2(x+1)}{x-2}$
10 $\frac{2(x+6)}{x-3}$

Solving quadratic equations (pp. 53–57)

1 By factorising

1 -2 or -3 **2** 5 or -2 **3** 7 or 3
4 0 or 9 **5** -3 **6** 6 or -6
7 4.5 or -4.5 **8** 8 or -3 **9** $\frac{1}{2}$ or -2
10 $\frac{1}{5}$ or $-\frac{2}{3}$ **11** $-\frac{2}{7}$ or $\frac{1}{3}$ **12** 2 or $\frac{-4}{3}$
13 -1 or $\frac{-4}{5}$ **14** -1 or $-\frac{1}{6}$

2 Using the quadratic formula, and
3 On a calculator

1 -3 or -4 **2** -1 or -3
3 2 or -8 **4** 4 or 1
5 -0.5 or -4 **6** 1.7071 or 0.2929
7 1.7863 or -1.1196 **8** 0.8968 or -2.2301
9 5.2361 or 0.7639 **10** 6.3723 or 0.6277
11 4.3028 or 0.6972 **12** 4.1623 or -2.1623

Solving quadratic inequations (p. 58)

1 $-2 < x < 6$ **2** $2 < x < 4$
3 $x \leq -7$ or $x \geq -2$ **4** $-1.8730 \leq x \leq 5.8730$
5 $x < 0.6972$ or $x > 4.3028$
6 $-6.7913 < x < -2.2087$

Solving rational quadratic equations (p. 59)

1 4.5 **2** 8
3 -1.819 or -4.183 **4** $x = 6$ or 7
5 $x = 3$ or $-0.\dot{6}$ **6** $x = 2$ or -0.2

Roots of equations (pp. 60–69)

1 Calculating the value of the discriminant and number of roots

1 $\Delta = 121$, 2 roots **2** $\Delta = 0$, 1 root
3 $\Delta = -31$, 0 roots **4** $\Delta = -7$, 0 roots
5 $\Delta = 25$, 2 roots **6** $\Delta = 0$, 1 root
7 $\Delta = 0$, 1 root **8** $\Delta = -39$, 0 roots
9 $\Delta = 7$, 2 roots **10** $\Delta = -532$, 0 roots
11 $\Delta = 400$, 2 roots **12** $\Delta = 0$, 1 root
13 $\Delta = 361$, 2 roots **14** $\Delta = 73$, 2 roots
15 $\Delta = 0$, 1 root **16** $\Delta = -700$, 0 roots
17 $\Delta = 0$, 1 root **18** $\Delta = -1360$, 0 roots

2 Finding values for *a*, *b* or *c* when given the number of roots

1 $p > 0.5625$ **2** $q = 4.5$
3 $p < 0.6944$ **4** $k = -21.\dot{3}$
5 $k < -2.0167$ **6** $q > 0.7813$
7 $p > 10.95545$ or $p < -10.95545$
8 $k < 1$ **9** $k < -10.0417$
10 $k = 1$
11 $d > 16.7980$ or $d < -2.7980$
12 $-3.5 < d < 0.5$
13 **a** $x = \frac{-(n-m) \pm \sqrt{(n-m)^2 - 12nm}}{2}$
b $(n - m)^2 - 12nm > 0$
14 **a** $x = \frac{-(n-m) \pm \sqrt{(n-m)^2 - 60nm}}{2}$
b $(n - m)^2 - 60nm < 0$

3 Solving when substitution is required

1 $p = 1 \Rightarrow x = -2$
2 $p = 3 \Rightarrow x = 12$ or $p = -1 \Rightarrow x = 4$
3 Either $p = -2 \Rightarrow \sqrt{x + 5} = -2$
No solution to this.
Or $p = 6 \Rightarrow \sqrt{x + 5} = 6$
$\therefore x = 31$

ISBN: 9780170354226

4 $p = 1 \Rightarrow x = 0$, or $p = -12$, but no values for x satisfy $3^x = -12$, so only one solution

5 Let $p = 4^x$. $p = 2 \Rightarrow x = 0.5$, or $p = -3$, but no values for x satisfy $4^x = -3$, so only one solution

6 $p = 4 \Rightarrow 2^x = 4$, so $x = 2$ or $p = 7 \Rightarrow 2^x = 7$, so $x = 2.8074$

7 $p = 4 \Rightarrow x^2 = 4$, so $x = \pm 2$, but $p = -2 \Rightarrow x^2 = -2$, which has no solution $\therefore$ Roots are ± 2

8 $p = 0.75 \Rightarrow x^2 = 0.75$, so $x = \pm 0.8660$ or $p = -0.5 \Rightarrow x^2 = -0.5$. This has no solutions.

Forming and solving quadratic equations (pp. 70–83)

1 Given information about the equation

a Finding unknown numbers

$x = 6$ or -2. Positive number $\Rightarrow x = 6$

1 He could be thinking of 1 or 9.

2 She could be thinking of 7 or -3.

3 $x = 8$, so his number is 16.

4 The two numbers are 11 and 16.

5 $x = 4 \Rightarrow$ numbers are 9 and 11.

6 $x = 7 \Rightarrow$ number is 14.

7 He thought of 11.

8 She is eight.

9 $x = 4 \Rightarrow$ number is 8.

10 Ages are 11, 13 and 15.

11 They are 12 and 27.

12 He thought of 3.

b Problems about lengths, areas and volumes

Height < 8 cm

1 Length < 11 cm 2 Height > 9 cm

3 Sides are 5 cm, 12 cm and 13 cm

4 Square has 5 cm edge, so rectangle is 16 cm x 8 cm

5 Dimensions are 5 cm x 7 cm x 8 cm

6 Width = 8.6 m 7 Width = 1.198 m

8 Width = 1.5 m 9 Width = 2.5 m

2 Given information about the graph

a $y = ax^2$ and $y = -ax^2$

1 $y = \frac{1}{2}x^2$ 2 $y = -2x^2$ 3 $y = -\frac{1}{4}x^2$

4 $y = \frac{1}{5}x^2$ 5 $y = -\frac{3}{2}x^2$ 6 $y = -5x^2$

b $y = ax^2 + c$

1 $y = \frac{3}{2}x^2 + 7$ 2 $y = -3x^2 + 2$

3 $y = -0.1x^2 - 10$ 4 $y = -0.2x^2 + 6$

5 $y = 0.25x^2 - 7$

c $y = a(x + d)(x - d)$

1 $y = 0.75(x + 4)(x - 4)$. Depth = 9 m

2 $y = -0.5(x + 4)(x - 4)$. Bar is 6 m high

3 $y = -0.04(x + 10)(x - 10)$. Bar is 10 m long

4 $y = 0.01(x + 15)(x - 15)$. Width of puddle is 10 m.

d $y = a(x - d)(x - e)$

1 $y = -0.25(x - 1)(x - 7)$. Height of string is 2 m.

2 $y = -0.2(x - 2)(x - 10)$. Length of string is 2 m.

3 $y = 0.05x(x - 16)$. Width of surface of water is 4 m.

4 $y = -0.005(x - 4)(x - 52)$. No. The maximum height of the ball is 2.88 m.

Rearranging equations (pp. 84–86)

1 Where the subject appears once

1 $r = \sqrt{\frac{A}{\pi}}$ 2 $b = \pm\sqrt{x + 4ac}$

3 $c = \frac{b^2 - x}{4a}$ 4 $x = \sqrt[3]{\frac{3S}{4\pi}}$

5 $L = \frac{dx}{n\lambda}$ 6 $y = \sqrt{\frac{2E}{k}}$

7 $V = \sqrt{\frac{rF}{m}}$ 8 $v = \frac{uf}{u - f}$

2 Where the subject appears more than once

1 $x = \frac{1}{y - 1}$ 2 $x = \frac{-y}{y - 3}$ or $\frac{y}{3 - y}$

3 $x = \frac{-2}{y - 3}$ or $\frac{2}{3 - y}$ 4 $x = \frac{2y + 1}{3y - 2}$

5 $x = \frac{-1 - 2y}{y - 1}$ or $\frac{2y + 1}{1 - y}$ 6 $x = \frac{3y}{y - 5}$

3 Where there is a root sign

1 $x = y^2 + 3$ 2 $x = \frac{y^2 - 6y + 10}{4}$

3 $x = y^2 - 4y + 7$ 4 $x = \frac{y^2}{1 - y^2}$

5 $x = \frac{2(2y^2 + 1)}{1 - y^2}$ 6 $k = \frac{4\pi^2 m}{T^2}$

7 $c = -x(ax + b)$ 8 $L = \frac{1}{4\pi^2 f^2 C}$

Practice questions (pp. 87–91)

Practice question one (pp 87 - 88)

a i $\frac{1}{3x^2}$ ii $8x^{3/4}$ or $8\sqrt[4]{x^3}$

iii $\sqrt{8x}$

b He got 3 Merit grades

c i $y = -2(x - 1)(x - 5)$ or $y = -2(x - 3)^2 + 8$ or $y = -2x^2 + 12x - 10$

ii Height = 6 m

Practice question two (pp 88 - 89)

a i $(2x + 3)(5x - 4)$

ii $x = -1.5$ or 0.8

b i $\frac{2(x + 3)}{(x + 1)}$

ii $x = -2$. $x = 3$ is not a solution because that would result in a divisor of 0

c $-1.3284 < k < 4.3284$

d $\sqrt{5x^2 - 5x + 10}$ or $\sqrt{5(x^2 - 2x + 2)}$

Practice question three (pp 88 - 89)

a i $x = 2$

ii $x = 32$

iii After 6.8433 years

iv The number of years (N) it takes for the population to increase te fold at an annual rate of increase of m percent.

b 3

c $x = -1$

d i $A = 2(1.40)^n$

ii 55 or 56 more possums

ISBN: 9780170354226